APPLYING STATISTICS IN THE COURTROOM

A New Approach
for Attorneys and
Expert Witnesses

APPLYING STATISTICS IN THE COURTROOM

A New Approach for Attorneys and Expert Witnesses

Phillip I. Good

CRC Press
Taylor & Francis Group
Boca Raton London New York

CRC Press is an imprint of the
Taylor & Francis Group, an **informa** business
A CHAPMAN & HALL BOOK

CRC Press
Taylor & Francis Group
6000 Broken Sound Parkway NW, Suite 300
Boca Raton, FL 33487-2742

First issued in paperback 2019

ISBN-13: 978-1-58488-271-8 (hbk)
ISBN-13: 978-0-367-39713-5 (pbk)

Library of Congress Cataloging-in-Publication Data

Good, Phillip I.
    Applying statistics in the courtroom : a new approach for attorneys and expert witnesses
/ Phillip I. Good.
        p. cm.
    Includes bibliographical references and index.
    ISBN 1-58488-271-9 (alk. paper)
    1. Forensic statistics—United States. I. Title.

KF8968.75 .G66 2001
347.73'67—dc21   2001028695

Catalog record is available from the Library of Congress

Visit the Taylor & Francis Web site at
http://www.taylorandfrancis.com

and the CRC Press Web site at
http://www.crcpress.com

Preface

For the rational study of the law ... the man of the future is the man of statistics and the master of economics.

Oliver Wendell Holmes, 1897

We have written this text for two audiences with a single common goal: to ensure attorneys and statisticians will work together successfully on the application of statistics in the law.

To be effective in the courtroom, a statistician must be able to think like a lawyer and present complex statistical concepts in terms a judge can understand. Thus, we present the principles of statistics and probability, not as a series of symbols, but in the words of jurists. Concepts take precedence over formula.

Statisticians who may be skeptical of this approach should remember *the law is what judges say it is.*

For the attorney, a change in point of view is also necessary. West has no categories labeled "cohort analysis" or "sample size." Sheppardize, but do not expect to find cases that uphold or reverse the statistical points raised. The opinions in this text are included for any or all of the following reasons:

- They provide concise definitions of statistical concepts.
- They provide lucid explanations of these concepts.
- They illustrate the presentation of statistical evidence.
- They reveal grounds on which a statistical argument may be successfully attacked.

Two further caveats: our coverage is not meant to be complete. If we write, "two other circuits have approved the use of statistical sampling," we do not mean to imply that other circuits have not approved it. Such statements imply only that our research has been limited. Finally, we have focused solely on the statistical issues in given cases. Many other issues are involved in almost every instance and, frequently, these issues, not the statistical ones, have formed the basis for a court's ultimate determination.

Hopefully, this text will serve the attorney as a comprehensive guide to the application of statistics and probability in jury selection, employment discrimination, trademark disputes, criminal law, civil law, and product liability. Sidebars and Chapters 13 and 15 indicate issues to raise during discovery and possible lines of counterattack, as well as potential areas of vulnerability.

This book is divided into four parts. The opening chapters concern the relationship between a sample and the population from which it is drawn. Chapter 1 describes the courts' gradual acceptance of samples and sampling methodology. Chapter 2 defines the representative random sample. Chapter 3 compares various sampling methodologies. Chapter 4 is devoted to the use of descriptive statistics in the courtroom including measures of central tendency, precision, and percentage.

Chapter 5 provides a brief introduction to probability. Chapters 6, 7, and 8 describe the varying acceptance of probability-based testimony in civil, criminal, and environmental hazard cases, respectively.

Chapter 9 summarizes the courts' responses to how large a sample must be. Chapter 10 addresses the same topic from the statistician's point of view and describes some simple procedures for testing statistical hypotheses. Chapter 11 describes correlation and regression of two variables. Chapter 12 extends this discussion to multiple variables and shows how the courts have applied multiple regression methods in cases of alleged discrimination.

Chapter 13 is devoted to preventive actions that can be taken to stay out of the courtroom and discusses the challenges that can be made to bad statistics once inside. Chapter 14 prepares the statistician for some of the twists and turns the trial process can take. Chapter 15 describes how an attorney can make the most effective use of statistics and statisticians at various points in the trial process and provides a set of questions on data-related concerns for use during discovery.

I would like to express my gratitude to the following libraries for the use of their facilities: the Orange County Law Library, the University of San Diego Law Library, and the Whittier Law School Library. Without these excellent community resources, this text would have been impossible.

Interpreting Case Citations

For those unfamiliar with legal research, here is a brief explanation of case citations such as *U.S. v. Two Bulls*, 918 F.2d 56, 61 (8th Cir. 1990). This citation refers to a decision handed down in 1990 by the court of appeals for the Eighth Circuit which embraces the Dakotas, Nebraska, Minnesota, Iowa, Missouri, and Arkansas. The appeal was brought by federal prosecutors (the U.S.) unhappy with the results of the earlier trial of Two Bulls (a Navaho Indian). This decision was reported in Volume 918 of the second series of the *Federal Reporter* beginning on page 56; the cited quotation will be found on page 61.

Prologue: An Introduction to the U.S. Court System for the Statistician

The purpose of this prologue is to describe for the statistician the court system in the U.S. and the roles the statistician will be expected to play.

The majority of the statistician's participation will take place behind the scenes as an advisor to the attorney. He or she may provide an estimate or a significance level and be asked to comment on and find weaknesses in the statistical procedures the attorney's opponent has used or is likely to use. The statistician's work product may appear in court as a portion of an attorney's brief (written motion), or it may never appear at all.

A trial is preceded by a series of pretrial motions and a process known as *discovery*, in which depositions are taken and the facts of the case are more or less agreed upon. The purpose of discovery is to reduce the time actually spent in the courtroom and, hopefully, to bring both parties to a settlement without having to hold a trial. Consider yourself a success if your efforts as a statistician contribute to a pretrial settlement.[1]

As a statistician, you may be asked to provide testimony under oath in the form of a *deposition*. This testimony will not be given in court and may be taken in an attorney's office or on neutral ground. You may be asked to testify by the attorney who hired you originally or by the attorney

---

[1] Assuming, of course, you don't persuade the attorney you work for that he or she has no case when, in fact, the case is an excellent one.

who is his or her opponent. In either case, you may be subject to a detailed (and unpleasant) examination or cross-examination by the opposing attorney. You can get revenge of a sort by counseling *your* attorney thoroughly so that he or she can subject the opposing party's statistician to an equally unpleasant afternoon of questions. In a third, relatively rare alternative, you may be summoned and employed by a judge to act as an impartial expert witness.

Many cases go to trial in a state or a federal court. During the trial, the judge may be asked to rule on various points of law that arise. An example would be to decide whether the results of a survey may be introduced in court. To guide his or her ruling, the judge may review rules that his or her own and higher courts have published, the published opinions of appeals and higher courts (*common law*), statutes (laws passed by legislatures), and other sources of information he or she considers helpful.

A ruling made by a federal district court may be reviewed by a federal circuit appeals court, and the appeals court ruling may be further reviewed by the U.S. Supreme Court. Not all cases are subject to review; the Supreme Court reviews only a handful of the appeals that come to it. A ruling by the Supreme Court is binding on all federal courts. A ruling by a circuit appeals court is binding only on the courts within its circuit, and its ruling is subordinate to whatever the Supreme Court later decides.

A ruling by a state court may be reviewed by a state appeals court and (depending on the state) may be reviewed subsequently by still higher level courts in that state. A ruling by the highest state court is binding on all the courts *within* that state, but it is not binding on the courts in other states.[2] Nor is a ruling in a federal court binding on any of the state courts.[3]

Decisions by state trial courts are almost never published and thus cannot be cited by attorneys. Some federal trial court opinions are published; most are not. Almost all appeals court and higher court opinions are published. Trial court judges look to those opinions for guidance, and we have looked to them for material on the application of statistics in the courtroom.

---

[2] Nonetheless, a judge in one state might well be interested in what the appeals courts in other states have decided if a ruling on a particular point of law has not yet been made in his or her own state.

[3] The exception is an appeal of a state court decision based on an alleged violation of federal law. An example would be the overturning by a federal court of a decision in a state that did not allow African-Americans to serve on its juries.

The Author

Phillip Good, Ph.D., a graduate of the University of California at Berkeley's program in mathematical statistics and a paralegal, is the author of 12 published books including *Permutation Tests* (Springer, 1994, second edition, 2000), and *Resampling Methods* (Birkhauser, 1999, second edition, 2001), as well as 34 scholarly publications, 12 published short stories, and over 600 popular articles. A former Calloway Professor of Computer Science at the University of Georgia, Division Head and Professor of Biology and Physics at West Coast University, and Associate Professor of Applied Mathematics at Claremont College, he has worked in the aerospace, computer, energy, marketing, pharmaceutical, and telecommunication industries. He has worked with attorneys on cases involving audits, jury selection, personal injury, and product liability.

Contents

PART IV: APPLYING STATISTICS IN THE COURTROOM

SAMPLES AND POPULATIONS

A fundamental principle of statistics is that a great deal may be learned about a population by taking a representative sample from it. A second, equally fundamental principle is that as the sample grows larger it will more closely resemble the population from which it is derived. Several questions arise immediately:

- Will the courts accept the use of a sample in place of a population? This is the principal topic of Chapter 1.
- How can we ensure a sample is representative? This is the topic of Chapter 2.
- How we can minimize sampling errors and reduce the cost of sampling is the topic of Chapter 3.
- In Chapter 4 we consider how best to present and summarize the sample for the court.

Discussion of the mathematically demanding topic of how large is *large* is deferred to Chapter 9.

Chapter 1

Samples and Populations

Early in this century, statisticians in industry and government began to make use of samples to describe and estimate the characteristics of the populations from which the samples were drawn. The new science of quality control led to contracts between manufacturers and suppliers incorporating provisions calling for the acceptance or rejection of an entire shipment based on tests of a small sample thereof.[1] At almost the same time, the courts began to recognize that representative random samples consisting of a few properly selected members of a population could take the place of the entirety.

In 1922, the government of the U.S., acting under the authority of the Pure Food and Drugs Act,[2] condemned an entire shipment of salmon, a total of 1974 cases, on the basis of two samples, each of 192 cans. In the first of these samples, 55 cans contained rotten and decayed salmon whose odor was offensive (putrid or tainted) or at the beginning of decomposition (stale). The second sample contained 47 putrid or tainted and stale cans. The lower court directed a verdict in favor of the manufacturer upon the ground that the article of food referred to in the statute was a single or individual can of salmon and not the entire case or lot. The Ninth Circuit appeals court reversed the verdict, holding that:

---

[1] The pioneering work on acceptance sampling by H.F. Dodge and H.C. Romig is described in Duncan [1986].

[2] 34 Stat. 768.

The word "article" is used in its broad and comprehensive sense, and has reference to the food product, not the smallest individual container. Any other construction would defeat the entire purpose of the law.

It upheld the act of condemnation, though it did say, somewhat tongue in cheek, that the manufacturer might cut its losses and separate the good salmon from the bad, providing "the burden of so doing should rest upon it, and not upon the government or the ultimate consumer. If it cannot do this, it is its own misfortune and it must suffer the consequence."[3]

In 1946, the FDA condemned a shipment of prophylactics as adulterated and misbranded after tests in which six of 82 of one brand examined in post-seizure tests, and eight of 108 of another examined pre-seizure contained holes. The manufacturer protested it was unfair to condemn the entire shipment; it also protested the tests because they were destructive in nature. The court noted the act referred specifically to samples and the need to make them available for testing and cited to *Andersen* before ruling in the FDA's favor.[4]

By the mid twentieth century, the use of samples in place of populations was well established in law in situations where the sampling process was inherently destructive. The extension to all samples and all populations was to follow. Today, the use of samples to detect welfare fraud and software copyright violation and to determine sentences in criminal cases is commonplace.

1.1 Audits

In *Ratanasen v. California Dept. of Health Services*,[5] the Ninth Circuit confronted a multitude of issues related to sampling:

- Can a random sample take the place of an exhaustive series of individual examinations?
- Are sampling and extrapolation properly questions of law?
- Are the appropriateness of sample size and other aspects of sample design questions of fact?
- What percentage of the population comprises an adequate sample?[6]

In *Ratanasen*, the Department of Health Services (DHS) audited claims for payments filed by Dr. Ratanasen, a physician, over a 2-year period,

[3] *Andersen & Co. v. U.S.*, 284 F. 542, 543 (9th Cir. 1922).
[4] *U.S. v. 43½ Gross Rubber Prophylactics*, 65 F. Supp. 534 (Minn. 4th Div. 1946).
[5] 11 F.3d 1467 (9th Cir. 1993).
[6] A further issue, the nature of the sampling process, is considered in Section 3.2.

and determined he had overbilled the Medi-Cal program $124,268. Subsequently, the state asked a bankruptcy court to award it this amount.

The bankruptcy court concluded the state's use of a random sample to determine the amount overbilled was valid and allowed the claim. The physician filed an objection on the grounds the claim was based on a random sample. The Ninth Circuit ruled the use of sampling and extrapolation as part of state audits in connection with Medicare and other similar programs is permissible to prove fraud, provided the aggrieved party has the opportunity to rebut such evidence.

DHS selected a sample of 300 Medi-Cal beneficiaries out of a total of 8761 beneficiaries for whom the physician had submitted claims during the period in question, and it used this sample to estimate the total overpayment. The physician argued that to determine the correct total of excess Medi-Cal payments, each file would have to be examined on its own. The bankruptcy court concluded that a creditor may prove the amount of its claim through the use of convincing statistical samplings. The bankruptcy court then held an evidentiary hearing and concluded that the method of statistical extrapolation used by the DHS was valid and in compliance with California law. In its Findings of Fact, the court stated:

> The Simple Random Sampling Method of statistical extrapolation utilized by the Department in calculating the liability of the Debtor pursuant to the audit was valid and in compliance with California law, Title 22, California Code of Regulations, 51458.2. The sample size, level of competence, and other measures of the extrapolation method used were appropriate and convincing ….

The physician maintained to the Ninth Circuit appeals court that this ruling contained conclusions of law or, at least, involved mixed issues of law and fact. The appeals court subsequently ruled that whether the use of sampling and extrapolation is proper is a question of law, while whether the sample size and other measures are appropriate are questions of fact.[7]

1.1.1 Validity of Using Sampling Methods

Whether the DHS could prove its claim by using statistical information is a matter of law. A Georgia district court addressed this issue[8] in a case involving the refusal of the Department of Health, Education and Welfare (HEW) to reimburse the state of Georgia for $3.5 million the state paid

---

[7] *Ratanasen v. California Dept. of Health Serv.*, 11 F.3d 1467, 1469 (9th Cir. 1993).

[8] *State of Georgia Dept. of Human Resources v. Califano*, 446 F. Supp. 404 (N.D. Ga. 1977).

to doctors who provided services to Georgia Medicaid recipients for three years in the 1970s. An audit by HEW, conducted on the basis of random statistical samples of claims paid during a five-quarter period, revealed that the state of Georgia had paid some claims in excess of ceilings imposed by federal statutes.[9] As a result, HEW disallowed $1.5 million in matching federal funds and made a demand for a refund of that money. The state appealed, claiming HEW's decision was arbitrary and capricious because the amount of overpayment was determined by use of a statistical sample rather than by an individual, claim-by-claim review. In finding for HEW, the district court concluded:

> [T]he use of statistical samples was not improper. Projection of the nature of a large population through review of a relatively small number of its components has been recognized as a valid audit technique and approved by federal courts in cases arising under Title IV of the Social Security Act. Moreover, mathematical and statistical methods are well recognized as reliable and acceptable evidence in determining adjudicative facts.[10]

The Seventh Circuit cited the preceding case in upholding the auditing procedures used by the state of Illinois in auditing physicians reimbursed with public funds for medical services.[11] Illinois audited a sample of 353 records randomly selected from a total of 1302 records for the audit period and determined a participating doctor had been overpaid $5018.[12] The doctor contended any formula for sampling and extrapolation was improper per se.[13] The court disagreed and concluded, "...the use of sampling and extrapolation is proper, provided there is an opportunity to rebut the initial determination of overpayment."[14]

The Sixth Circuit also cited *State of Georgia Dept. of Human Resources v. Califano* in a case where the Department of Education found as a result of an audit conducted through *random, stratified sampling*[15] that the Michigan Department of Education misspent federal funds in the conduct of its vocational rehabilitation program.[16] The court ruled that audits of thousands of cases "comprising the universe of cases" would be impossible.[17] Final determination of the total invalid expenditures was not made

---

[9] Id. at 406.

[10] Id. at 409 (citations omitted).

[11] *Illinois Physicians Union v. Miller*, 675 F.2d 151, 156 (7th Cir. 1982).

[12] Id. at 152.

[13] Id. at 155.

[14] Id. at 156.

[15] See Section 3.2 for a discussion of stratified sampling.

[16] *Michigan Dept. of Educ. v. U.S. Dept. of Educ.*, 875 F.2d 1196 (6th Cir. 1989).

[17] Id. at 1205.

until the state had a chance to present its own evidence of an error in the audit.

The Second, Sixth, and Seventh Circuits have approved the use of statistical sampling for welfare fraud and medical reimbursements. In *Chaves County Home Health Serv., Inc. v. Sullivan*,[18] the District of Columbia Circuit evaluated the contention of home health care providers that the Secretary of Health and Human Services (HHS) had improperly suspended the existing individual claims adjudication process under the Medicare Act and replaced it with a method based on statistical sampling to determine overpayments. The court cited the three rulings already discussed here in affirming summary judgment for HHS, stating:

> [W]e agree with HHS that the statutory scheme of individualized review of claims on pre-payment review can be reconciled with a sample adjudication procedure on post-payment review. Such an interpretation is reasonable given the logistical imperatives recognized by courts in other comparable circumstances.[19]

In *Ratanasen*, the Ninth Circuit joined the other circuits in approving the use of sampling and extrapolation as part of audits in connection with Medicare and similar programs.[20] The court stated:

> To deny public agencies the use of statistical and mathematical audit methods would be to deny them an effective means of detecting abuses in the use of public funds. Public officials are responsible for overseeing the expenditure of our increasingly scarce public resources and we must give them appropriate tools to carry out that charge.

1.1.2 Basis for Objection

Although the courts have routinely upheld the use of random sampling by federal and state agencies in recoupment actions,[21] providers have

---

[18] 931 F.2d 914 (D.C. Cir. 1991), cert. denied, 502 U.S. 1091 (1992).

[19] Id. at 919.

[20] See also *Mercy Hospital of Watertown v. New York State Dept. of Social Services*, 79 N.Y.2d 197, 590 N.E.2d 213, 581 N.Y.S.2d 628 (1992).

[21] *Yorktown Medical Laboratory, Inc. v. Perales*, 948 F.2d 84 (2nd Cir. 1991) (sampling does not violate due process); *Mile High Therapy Centers Inc. v. Bowen*, 735 F. Supp. 984, 986 (D.D.C. 1988) (Secretary has authority under 42 U.S.C. 1395u(a) to establish rules for sampling); *Chaves County Home Health Service v. Sullivan*, 931 F.2d 914, 922 (D.C.C. 1991) (requiring agency to engage in analysis of every claim represents a "daunting burden"); *Protestant Memorial Medical Center, Inc. v. Dept. of Public Aid*, 295 Ill. App. 3d 249, 251 (1998).

successfully attacked random sampling by demonstrating that (1) the sample was not random;[22] (2) the sample size was too small;[23] or (3) a stratified sample should have been used.[24] (See also Bierig, 1998.)

Whether a sample is truly random can be determined by reference to statute (Chapter 2) or the laws of probability (Chapter 5). The choice of sample size is considered in the next section and again in Chapter 9. The appropriate use of stratified sampling is considered in Section 3.2.

1.1.3   Is the Sample Size Adequate?

In *Ratanasen,* the physician contended the sample size of 3.4% of the population was so small as to represent a violation of due process. He asked the Ninth Circuit to consider *Daytona Beach General Hospital, Inc. v. Weinberger,*[25] in which a recoupment due from a hospital because of alleged overpayments was calculated by using a sampling method based on less than 10% of the total cases in question. In that case, the district court found that the 10% procedure denied the plaintiff due process and remanded the case for the Secretary of Health, Education, and Welfare to review.[26]

The Ninth Circuit found no basis in the cases cited by the physician for a statistical "floor" that auditors must exceed in order to guarantee providers due process.[27] The number of items in the *Ratanasen* sample of 3.4% exceeded the number of items in the sample in *Michigan Dept. of Education v. U.S. Dept. of Education,*[28] in which a random, stratified sample of 4% was used as a starting point for determining improper expenditures.

The Sixth Circuit noted in the latter case, "There is no case law that states how large a percentage of the entire universe must be sampled."[29] This is not quite true, and we return to the choice of a sufficiently large sample size in Chapter 9.

1.2   Determining the Appropriate Population

The courts and statutory law state that a sample must be drawn from a population germane to the issue under adjudication. We illustrate such

---

[22] *U.S. v. Skodnek,* 933 F. Supp. 1108 (D. Mass. 1996).

[23] *Fisher v. Vassar College,* 70 F.3d 1420, 1451 (2nd Cir. 1995).

[24] *Protestant Memorial Medical Center, Inc. v. Dept. of Public Aid,* 692 N.E.2d 861 (Ill. App. 1998).

[25] 435 F. Supp. 891 (M.D. Fla. 1977).

[26] Id. at 900.

[27] 11 F.3d 1467 at 1472 (9th Cir. 1993); see also Chapter 8.

[28] 875 F.2d 1196, 1199 (6th Cir. 1989).

[29] Id. at 1199.

frames development in the following sections with opinions from jury panel selection, criminal law, trademark law, and discrimination cases.

1.2.1 Jury Panels

The Sixth Amendment provides that:

> A criminal defendant is entitled to a jury drawn from a jury panel which includes jurors residing in the geographic area where the alleged crime occurred.

The phrase "geographic area" is ambiguous, and the courts have struggled to create an exact definition. *Taylor v. Louisiana*[30] held that juries shall be drawn from "the state and district wherein the crime shall have been committed."

The vicinage in a criminal prosecution can be greater in size than the venue or place of trial.[31] "For purposes of the fair cross-section jury requirement, the judicial district, and not the county nor a 20-mile courthouse radius, constitutes the community. However, for the purposes of vicinage, a right discrete from that to a representative jury, the county is the appropriate measure."[32]

In *People v. Sirhan*,[33] the California Supreme Court rejected a contention of bias because the anecdotal evidence offered by the defense based on the population of the northern U.S. was unrelated to the trial court venue (Los Angeles County).

In *People v. Harris*,[34] this same court ruled that a survey of trial court jury panels showed a significant disparity from supplemented census figures and thus was a basis for reversal. Responding to the prosecution's contention that the survey was too limited, being restricted to the superior courts in a single district, Justice Mosk noted the Balkanized nature of Los Angeles County and referred to the "significant deceptiveness" of countywide statistical data.[35]

---

[30] 419 U.S. 522 (1975).

[31] *People v. Flores*, 133 Cal. Rptr. 759, 62 Cal. App. 3d Supp. 19 (1976). See also *Adams v. Superior Court of Los Angeles County*, 27 Cal. App. 3d 719, 104 Cal. Rptr. 144 (1972).

[32] *California v. Harmon*, 215 Cal. App. 3d 552, 263 Cal. Rptr. 623 (1989).

[33] 7 Cal. 3d 710, 102 Cal. Rptr. 385 (1972), cert. denied, 410 U.S. 947.

[34] 36 Cal. 3d 36, 201 Cal. Rptr. 782 (1984), cert. denied, 469 U.S. 965, appeal to remand, 236 Cal. Rptr. 680, 191 Cal. App. 3d 819, appeal after remand, 236 Cal. Rptr. 563, 217 Cal. App. 3d 1332.

[35] The frame or source population must be kept current. See Section 2.3.2.

1.2.2 Criminal Universe

Traces of blood have been found at the crime scene; a sample of the defendant's blood is taken, and DNA fragments from the two sources match. What is the likelihood that DNA fragments taken from some other individual also would have matched?

Statisticians and jurists agree the answer depends on the population from which this other random individual is drawn. For example, if the defendant is an American Indian and a substantial part of the suspect population shares the defendant's heritage, then the FBI's DNA database drawn from a demographically diverse but predominantly Caucasian group is *not* the appropriate basis of comparison.[36]

Lempert [1994] proposes that the appropriate population of suspects be determined by examining the other evidence of a crime. Suppose a rape victim lives in an inner city ghetto, predominantly populated by African-Americans. Our attention would turn to DNA fragments taken from African-American males. But if the rape victim says her assailant is white, perhaps the FBI's predominantly white database is appropriate after all.[37]

1.2.3 Trademarks

Amstar Corporation claimed that Domino's Pizza was too easily confused with Amstar's use of the "Domino" trademark for sugar. It brought suit alleging trademark infringement, unfair competition, and dilution. The district court entered judgment in favor of the plaintiff, dismissing the defendant's cross-complaint, and defendants appealed. The appeals court reversed, in part because surveys both parties used to support their claims were substantially defective.[38]

> A finding of fact is clearly erroneous "when although there is evidence to support it, the reviewing court on the entire evidence is left with the definite and firm conviction that a mistake has been committed."[39] In undertaking to demonstrate likelihood of confusion in a trademark infringement case by use of

---

[36] See *U.S. v. Two Bulls*, 918 F.2d 56 (8th Cir. 1990); *State v. Passino*, No 185-1-90 Fcr (Dist. Ct. Franklin County May 13, 1991).

[37] See also Coleman and Walls [1974], Smith and Charrow [1975], and Lempert [1991].

[38] *Amstar Corp. v. Domino's Pizza, Inc.*, 205 U.S.P.Q 128 (N.D. Ga. 1979), rev'd, 615 F.2d 252 (5th Cir. 1980).

[39] Ibid. citing at page 258 to *U.S. v. U.S. Gypsum Co.*, 333 U.S. 364, 395 (1948) and to *Handbook of Recommended Procedures for the Trial of Protracted Cases*, 25 F.R.D. 351, 429 (1960).

survey evidence, the appropriate universe [or frame] should include a fair sampling of those purchasers most likely to partake of the alleged infringer's goods or service.[40]

Amstar conducted and offered in evidence a survey of heads of households in ten cities. Domino's Pizza had no stores or restaurants in eight of these cities. Its outlets in the remaining two cities had been open less than three months. Only women were interviewed by Amstar. Those women were at home during daylight hours; that is, they were grocery shoppers rather than young and single women who comprise the majority of pizza eaters. Similarly, the court rejected Domino Pizza's own survey conducted in its pizza parlors. Neither plaintiff nor defendant had sampled from a sufficiently complete universe.

Courts are more likely to respond positively to carefully focused survey results. In *Windsurfing Int'l v. Fred Osterman GMBH*,[41] defendants were charged with unauthorized use of the "Windsurfer" trademark. They responded that *windsurfer* had become a generic term and offered in evidence a survey they had conducted in which those surveyed were asked whether each of eight product names was a brand name or a common name. To eliminate guessing, respondents were also asked what kinds of products they were asked about. The survey indicated that 58% did not know what a "windsurfer" was. Of the knowledgeable 42%, 61% believed that *windsurfer* was merely a type of product. The court agreed and ruled that the word had become generic. The results of the complete survey are shown in Table 1.1.

Table 1.1   Results of the Windsurfer Test

Product	Brand	Common Name	Don't Know
STP	85	15	
Thermos	44	53	3
Margarine	10	90	
Windsurfer	36	61	3
Jello	71	27	2
Refrigerator	6	94	
Aspirin	11	89	

[40] Id. citing to *American Basketball Ass'n v. AMF Voit, Inc.*, 358 F. Supp. 981, 986 (S.D. N.Y.), aff'd, 487 F.2d 1393 (2nd Cir. 1973), cert. denied, 416 U.S. 986 (1974) and *Hawley Products Co. v. U.S. Trunk Co.*, 259 F.2d 69, 77 (1st Cir. 1958).

[41] 613 F. Supp. 933 (S.D. N.Y. 1985)

1.2.4 Discrimination

Selecting the appropriate and relevant population for comparison is essential.

In *Hazelwood School District v. U.S.*,[42] the federal government alleged that the Hazelwood School District, located in St. Louis County, Missouri, and various officials, engaged in a "pattern or practice" of teacher employment discrimination in violation of Title VII of the Civil Rights Act of 1964.

The court considered whether the statistical data proffered by the federal government (the plaintiffs) appropriately supported its allegations. The plaintiffs had compared the proportions of teachers and pupils in Hazelwood who were African-American. The district court, following trial, ruled that the government had failed to establish a pattern or practice of discrimination.[43] The Eighth Circuit court of appeals reversed, in part on the ground that the trial court's analysis of statistical data should have entailed a comparison between the racial composition of Hazelwood's teaching staff and the racial composition of the qualified public school teacher population in the relevant labor market.[44]

In the 1972–1973 and 1973–1974 school years, only 1.4% and 1.8%, respectively, of Hazelwood's teachers were African-Americans.[45] This statistical disparity, particularly when viewed against the background of Hazelwood's teacher hiring procedures, was held to constitute a *prima facie* case of a pattern or practice of racial discrimination.[46] For many years, Hazelwood followed relatively unstructured procedures in hiring its teachers. Each school principal had virtually unlimited discretion in hiring teachers for his or her school. Hazelwood hired its first African-American teacher in 1969. The number of African-American faculty members gradually increased in successive years: 6 of 957 in 1970; 16 of 1107 by the end of the 1972 school year; and 22 of 1231 in 1973. By comparison, according to 1970 census figures, of more than 19,000 teachers employed in that year in the St. Louis area, 15.4% were African-American. That percentage figure included the St. Louis City School District, which in recent years had followed a policy of attempting to maintain a 50% African-American teaching staff. Apart from that school district, 5.7% of the teachers in the county were African-American in 1970.[47]

Selecting St. Louis County and the city of St. Louis as the relevant area, the appeals court compared the 1970 census figures, showing that 15.4% of teachers

---

[42] 392 F. Supp. 1276 (E.D. MO), rev'd, 534 F.2d 805, vacated and remanded, 433 U.S. 299 (1977).

[43] Ibid.

[44] 534 F.2d 805 as stated at 433 U.S. 299, 305 (1977).

[45] 433 U.S. 299, 301–302.

[46] Id. at 305.

[47] Id. at 301–302.

in that area were African-American compared to the racial composition of Hazelwood's teaching staff. In the 1972–1973 and 1973–1974 school years, only 1.4% and 1.8%, respectively, of Hazelwood's teachers were African-Americans.[47]

Preventive Statistics

The ruling by the appeals court in *Hazelwood* is not quite the end of the story. As the Supreme Court noted in *Teamsters v. U.S.*:

> Statistics are not irrefutable; they come in infinite variety and, like any other kind of evidence, they may be rebutted. In short, their usefulness depends on all of the surrounding facts and circumstances.[48]

In reviewing the lower court findings in *Hazelwood,* the Supreme Court stated,

> The role of statistics in 'pattern or practice' suits under Title VII provides substantial guidance in evaluating the arguments advanced by the petitioners. In that case, we stated that it is the Government's burden to establish by a preponderance of the evidence that racial discrimination was the [employer's] standard operating procedure — the regular, rather than the unusual, practice.[49] We also noted that statistics can be an important source of proof in employment discrimination cases, since, absent explanation, it is ordinarily to be expected that nondiscriminatory hiring practices will, in time, result in a workforce more or less representative of the racial and ethnic composition of the population in the community from which employees are hired. Evidence of long-lasting and gross disparity between the composition of a workforce and that of the general population thus may be significant even though section 703(j) makes clear that Title VII imposes no requirement that a workforce mirror the general population.[50] Where gross statistical disparities can be shown, they alone may, in a proper case, constitute *prima facie* proof of a pattern or practice of discrimination.[51]

---

[48] 431 U.S. 324 at 340 (1977).

[49] Id. at 336.

[50] Id. at 336. See also *Arlington Heights v. Metropolitan Housing Dev. Corp.*, 429 U.S. 252, 266 (1977); *Washington v. Davis*, 426 U.S. 229, 241-242 (1976).

[51] *Hazelwood* at 308; *Teamsters* at 339.

In light of *Teamsters*, the District Court's comparison of Hazelwood's teacher workforce to its student population fundamentally misconceived the role of statistics in employment discrimination cases. The court of appeals was correct in the view that a proper comparison was between the racial composition of Hazelwood's teaching staff and the racial composition of the qualified public school teacher population in the relevant labor market.[52]

Pre-Act versus Post-Act Discrimination

While upholding the general principles enunciated in this chapter, the Supreme Court vacated the judgment of the court of appeals on the grounds that a defendant employer must be given an opportunity to show that "the claimed discriminatory pattern is a product of pre-Act hiring, rather than unlawful post-Act discrimination."[53]

For the 1972–1973 school year, Hazelwood hired 282 new teachers, 10 of whom (3.5%) were African-Americans; for the following school year, it hired 123 new teachers, 5 of whom (4.1%) were African-Americans. Over the two-year period, African-Americans constituted a total of 15 of the 405 new teachers hired (3.7%).

> Although the Court of Appeals briefly mentioned these data in reciting the facts, it wholly ignored them in discussing whether the Government had shown a pattern or practice of discrimination. And it gave no consideration at all to the possibility that post-Act data as to the number of African-Americans hired compared to the total number of African-American applicants might tell a totally different story.[54]

We return to a study of the use of statistics in discrimination cases in Chapter 12.

Related Cases

In *Ward's Cove Packing Co. v. Antonio*,[55] the court specified the proper comparison was between the proportion of minority workers seeking the jobs at issue and the proportion in the area workforce. Yet in the more recent *McNamara v. City of Chicago*,[56] the court let stand a 7th District

---

[52] See *Teamsters* at 337–338, and n. 17.

[53] 431 U.S. 310 at 360.

[54] Id. at 310.

[55] 490 U.S. 642 (1989).

[56] 138 F.3d 1219 (7th Cir. 1998), cert. denied, 520 U.S. 981, (1998).

appeal court ruling that allowed a comparison with the minority proportion in the community at large.

1.2.5 Downsizing

One instance that presents substantial difficulty in determining the correct population for comparative purposes occurs during downsizing. If only one office is closed, as both parties agreed in *Parcinski v. Outlet Co.*,[57] then the correct comparison is between those transferred and those terminated; but if two offices are consolidated as the appeals court ruled in *Marisco v. Evans Chemetics*,[58] then those terminated should be compared with those retained.

This distinction can lead to contradictory rulings. In *Marisco*, the age distribution of those transferred (three of four were in the protected class) was approximately the same as the distribution of those terminated. Arguing, as Evans did, that only a single office was closed, no inference of discrimination arises. The district court supported this view and ruled for the defendant.

Marisco argued the two offices were consolidated, and that accounting employees should be distinguished from nonaccounting personnel. Twenty-three accounting personnel were at the two offices before the consolidation, of whom three were in the protected class; there were 20 after the consolidation, none of whom were in the protected class — clear evidence of discrimination. The Second Circuit appeals court reversed and remanded, supporting this latter view.

1.3 Summary

The use of sampling and extrapolation is well established under law. Whether the sample size and other measures are appropriate are questions of fact:

1. The sample must be drawn from a population germane to the issue under adjudication; testimony from one or more domain experts would be helpful in making such a determination.
2. The sampling method must be appropriate and lawful; see Chapter 2 and Section 3.2.
3. The sample size must be adequate; see Chapter 9.

---

[57] 673 F.2d 34 (2nd Cir. 1982), <u>cert. denied,</u> 459 U.S. 1103 (1983).
[58] 964 F.2d 106 (2nd Cir. 1992).

Chapter 2

Representative Samples and Jury Selection

2.1 Concepts

To obtain a random, representative sample, each individual (or item) in a population must have an equal probability of being selected. No individual (or item) or class of individuals may be discriminated against. As almost all our readers will have had some experience with the jury system,[1] we illustrate the drawing of a random, representative sample through a study of the process of jury selection. For completeness, the fundamental issues in jury selection including the right to serve, the basis on which jury selection may be challenged, and the right to challenge are also presented.

2.2 Issues

At issue is whether a jury verdict can be challenged successfully on the basis of its demographic composition or the demographics of the panel from which it was selected. The answer is affirmative if a *prima facie* case can be established based on one of the following criteria:

---

[1] I received a summons to appear for jury duty while writing this section.

- The selection process discriminated against jurors because of their race, gender, religion, party affiliation, or membership in some other distinguishable group that the court can be persuaded to recognize.
- Absent such discrimination, the statutory selection process incorporating random representative selection was not adhered to.

The composition of any specific panel is irrelevant. The demonstrated bias or error must be in the process of selection.

2.2.1 Burden of Proof

The party contesting the composition of the jury bears the initial burden of proof that discrimination exists. "A claim of denial of due process right requires a showing that the jury selection process tended to exclude or underrepresent some discernible class of persons."[2] The petitioner must show three things:

1. The group alleged to be excluded is a distinctive, cognizable group (see Section 2.2.3).
2. Representation in this group in venires from which juries are selected is not fair and reasonable in relation to the number of such persons in the community.[3]
3. The underrepresentation is due to systematic exclusion of the group in the juror selection process.[4]

In criminal cases, the burden of proof of discrimination can be shifted from the defendant to the state if the defendant can show either a *prima facie* case or strong inference of discrimination (such as no members of a given class known to have served on juries over an extended period of time). "Because of difficulty in obtaining more accurate figures for jury eligibility, defendant can present a *prima facie* case of violation of his right to a jury drawn from a representative cross-section of the community by showing through population figures a significant underrepresentation of a cognizable class, whereupon burden shifts to the state to demonstrate the absence of discrimination."[5]

---

[2] *U.S. v. Kennedy*, 548 F.2d 608, 614 (5th Cir. 1977).
[3] Here is where the statistician is so often called as an expert witness.
[4] See, for example, *Duran v. Missouri*, 439 U.S. 357 (1979); *People v. Harmon*, 215 Cal. App. 3d 552, 263 Cal. Rptr. 623 (1989).
[5] *People v. Harris*, 36 Cal. 3d 36, 201 Cal. Rptr. 782 (1984), cert. denied, 469 U.S. 965, appeal to remand, 191 Cal. App. 3d 819, 236 Cal. Rptr. 680, appeal after remand, 217 Cal. App. 3d 1332, 236, Cal. Rptr. 563.

2.2.2 The Right to Be Eligible to Serve

In the United States of America, the Declaration of Independence states, "All men are created equal." To ensure a jury of our peers, we need only ensure that all those eligible for jury duty do in fact have opportunities to be selected. The Federal Jury Selection and Service Act of 1968 as revised[6] states that citizens cannot be disqualified from jury duty "on account of race, color, religion, sex, national origin or economic status."[7] Subsequent rulings have upheld this section and extended its provisions to discrimination based on party affiliation.

Each U.S. district court is responsible for creating a written plan as to the operation of the jury system that meets the requirements of the Federal Jury Selection and Service Act.[8]

State laws creating discriminatory qualifications are not binding and cannot be enforced.[9] A state law that specifies only "white male persons shall be eligible to serve as jurors" is a violation of rights granted by the Fourteenth Amendment.[10] A federal jury panel from which women are intentionally and systematically excluded is not properly constituted.[11]

In *Patton v. Mississippi*,[12] the Supreme Court held a state cannot deprive a class of citizens the right to serve on a jury, either by statute or by administrative practices.

For more than a century, the Supreme Court in an unbroken line of cases (and only one dissent) has held that "A criminal conviction of a Negro cannot stand under the Equal Protection Clause of the Fourteenth Amendment if it is based on an indictment of a grand jury from which Negroes were excluded by reason of their race."[13] In *Rose v. Mitchell*,[14] the court set aside a conviction because of such exclusion and ordered the indictment quashed, notwithstanding that no constitutional impropriety had tainted the selection of the petit jury and guilt had been established beyond a reasonable doubt at a trial, free from constitutional error.

State laws concerning jury selection tend to mirror the federal laws. The California Trial Jury Selection and Management Act,[15] for example, disallows "exemption by reason of occupation, race, color, religion, sex,

---

[6] 28 U.S.C.A. §1861 et. seq. (1993).

[7] See 28 U.S.C.A. §1862 (1993).

[8] See 28 U.S.C.A. §1863 (1993).

[9] *Kie v. U.S.*, 27 Fed. 351, 357 (C.C. Ore. 1886).

[10] *Strader v. West Virginia*, 100 U.S. 303, 306 (1879).

[11] *Ballard v. U.S.*, 329 U.S. 187, 193 (1946); see also Babcock [1993].

[12] 332 U.S. 463.

[13] *Alexander v. Louisiana*, 405 U.S. 625, 628 (1972); see also *Bush v. Kentucky*, 107 U.S. 110, 119 (1883); *Neal v. Delaware*, 103 U.S. 370, 394 (1881).

[14] 443 U.S. 545, 550-564 (1979).

[15] Title 3, C.C.P. 204.

national origin, or economic status, or for any other reason." California courts have consistently ruled that potential jurors may not be barred from jury service on the basis of race, gender, or religion,[16] while permitting jurors to be excused for undue hardship.

2.2.3 Cognizable, Separate, Identifiable Groups

The jury selection process can be challenged if it can be shown that some separate, distinguishable group is discriminated against.[17] For example, black women constitute a cognizable group,[18] as do Asians.[19] A cognizable group may be established through evidence of the community's attitude toward members of the group; in *Hernandez v. Texas*,[20] a survey of community attitudes was used to demonstrate that Hispanics constitute a cognizable group.

Not every group will qualify for consideration; examples of groups that do not qualify include:

- Jurors fervently opposed to the death penalty[21]
- Jurors impartial to the underlying crime[22]
- Ex-felons and resident aliens[23]
- Jurors not possessing sufficient knowledge of English[24]
- The "less educated," (12 or fewer years of formal education) or "blue collar workers"[25]

In *Thiel v. Southern Pacific Co.*, the Supreme Court ruled that, "The exclusion of all those who earn a daily wage cannot be justified by federal or state law."[26] The Fifth Circuit court was willing to take cognizance of food stamp recipients while ruling they were not underrepresented.[27]

The door remains open for a showing that other potentially excluded groups such as young people, Italian-Americans, the deaf, the visually disabled, or single mothers constitute distinctive groups in the community.

---

[16] *People v. Fields*, 35 Cal.3d 329, 197 Cal. Rptr. 803 (1983).
[17] Statistical criteria for establishing such discrimination are given in Chapters 10 and 11.
[18] *People v. Motton*, 39 Cal.3d 596, 217 Cal. Rptr. 771 (1985), rehearing denied.
[19] *U.S. v. Cannady*, 54 F.3d 544 (9th Cir. 1995), cert. denied, 116 S.Ct. 210.
[20] 347 U.S. 475 (1954); see also *U.S. v. Rodriguez*, 588 F.2d 1000 (5th Cir. 1979).
[21] *Lockhart v. McCree*, 476 U.S. 162, 175 (1980).
[22] *U.S. v. Johnson*, 990 F.2d. 1129 (9th Cir. 1993).
[23] *People v. Pride*, 3 Cal. 4th 195, 10 Cal. Rptr. 2d 636 (1992), as modified on denial of rehearing.
[24] *People v. Lesara*, 206 Cal. App. 3d 1305, 254 Cal. Rptr. 417 (1988).
[25] *People v. Estrada*, 93 Cal. App. 3d 76, 155 Cal. Rptr. 731 (1979).
[26] 1328 U.S. 217 (1946).
[27] *U.S. v. Goff*, 509 F.2d 825 (5th Cir. 1975), cert. denied, 423 U.S. 857.

2.2.4 Voir Dire *Rights of Litigants and Jurors*

The right not to be discriminated against applies to all phases of the jury selection process including *voir dire.*

In federal criminal cases, the prosecutor may not exercise peremptory challenges based on race. In *Batson v. Kentucky,*[28] the Supreme Court remanded a criminal case for a determination of the reason the prosecutor used his peremptory challenges to strike all four blacks from the venire from which the jury, which convicted James Batson, an African-American, of burglary and receipt of stolen goods, was selected. This ruling of law was reaffirmed in *Powers v. Ohio*[29] and again in *Georgia v. McCollum,*[30] though in the latter case it was a group of white defendants charged with assaulting African-Americans that directed peremptory challenges based solely on the race of potential jurors.

Peremptory challenges by private parties in civil cases based solely on discriminatory factors are also prohibited.[31]

In *U.S. v. De Gross,*[32] the Ninth Circuit held that the Fifth Amendment's equal protection principles prevented gender-based peremptory strikes of venire persons by either the prosecution or the federal criminal defendant. Similarly, in *J.E.B. v. Alabama ex rel T.B.,*[33] the court held the state's use of peremptory challenges to exclude males from the jury was improper. "Gender like race is an unconstitutional proxy for juror competence and impartiality."

Preventive Statistics: Sample Size

In *Edmonson v. Leesville Concrete,* peremptory challenges were issued to only two of three potential African-American jurors. Is this statistically significant? And what if three of seven black venire persons were struck compared to three of 21 whites as happened in *U.S. v. Jordan,* 893 F.2d 182 (8th Cir, 1990)? The effect of sample size in jury selection and other matters is considered in Chapter 9.

---

[28] 476 U.S. 79 (1980).

[29] 499 U.S. 400 (1991).

[30] 505 U.S. 42 (1992).

[31] *Edmonson v. Leesville Concrete Co.,* 500 U.S. 614 (1991).

[32] 960 F.2d 1433 (9th Cir. 1992).

[33] 511 U.S. 127 (1994).

2.3 Composition of the Jury Pool

An appeal may also be based on the composition of the pool from which prospective jurors were selected.[34] The Sixth Amendment states that a "criminal defendant has right to trial by impartial jury drawn from a representative cross-section of the community." The Fourteenth Amendment extends this Sixth Amendment requirement to state courts in criminal proceedings.[35] The constitutions of many of the states contain similar language[36] and have also been used as a basis for reversal where the composition of the jury pool was improper.[37]

2.3.1 True Cross-Section

The Jury Selection and Service Act of 1968 states that all litigants entitled to jury trial shall have the right to a jury "selected at random[38] from a fair cross-section of the community in the district or division where the court convenes."[39] Most states have enacted regulations designed to achieve this end. For example, the *California Code of Civil Procedure*, Section 197 provides that a:

> [L]ist of registered voters and the Department of Motor Vehicles list of licensed drivers and identification card holders ... shall be considered inclusive of a representative cross section of the population.

In addition to voter registration and motor vehicle records, sources of jurors may include customer mailing lists, telephone directories, and utility company lists.[40] The use of combined lists of registered voters and licensed drivers is permitted.[41]

---

[34] See Section 1.2.1.
[35] 28 U.S.C.A. §1861 (1993); see also *Duncan v. Louisiana*, 391 U.S. 145, 149 (1968) (a criminal case); *Ristaino v. Ross*, 424 U.S. 589 (1976) (a civil case).
[36] See, for example, the *California Constitution*, Article I, Section 16.
[37] See, for example, *People v. Wheeler* (1979).
[38] The term *random* is defined explicitly in Section 2.4 and Chapter 5.
[39] *Duran v. Missouri*, 439 U.S. 357 (1979).
[40] Ibid.; but see *People v. White*, 43 Cal.2d 740 (1954), as well as an earlier *obiter dicta* in *Glasser v. U.S.*, 315 U.S. 60 (1942), ruling that the membership rosters of private clubs are too restrictive.
[41] *U.S. v. Bailey*, 862 F. Supp. 277 (D. Colo. 1994), aff'd in part, rev'd in part, 76 F.3d 320, cert. denied, 116 S.Ct. 1889.

Some quarrel with these provisions, noting that, "Potential jurors with … higher income, higher education … [are] more likely to be represented on juries because they are more inclined to register to vote …."[42]

California courts have taken the position that the use of voter registration lists as the sole source of jurors is constitutionally valid absent a showing that use of these lists results in systematic exclusion of a cognizable group. A federal court held that a defendant's contention that Hispanics were less likely to return questionnaires used in selecting jury panels and less likely to vote did not constitute systematic exclusion.[43]

2.3.2 A Snapshot in Time

Because the U.S. continues to be a magnet for the oppressed of all nations and because shifting technologies require an increasingly mobile work force, a jury panel that is representative of a population at the moment it is gathered may not be as representative a few years later. The courts have held that jury panels selected at least once every four years are adequate and need not be updated continuously.[44]

2.3.3 Composition of the Individual Panel

Variation is inherent in any sampling procedure. Consequently, the courts distinguish between a procedure and its outcome in any specific instance. For example, the demographic composition of a jury selected at random will vary from case to case. While members of an identifiable group may not be systematically excluded from a jury, they need not be present in any actual panel.

In *People v. Manson*,[45] the court held the defendants were not entitled to a jury of any particular composition, nor was there a requirement that the petit jury be representative of various distinct economic, political, social, or racial groups in the community.

In *Thiel v. Southern Pacific*, the court held it irrelevant that the jury in the case at issue contained five members of the excluded class:

---

[42] Fukurai et al. [1991]; see also Carp [1982].

[43] *U.S. v. Ortiz*, 897 F. Supp. 199 (E.D. Pa. 1995). See also *U.S. v. Lewis*, 472 F.2d 252, 255 (3rd Cir. 1973); *U.S. v. Cecil*, 836 F.2d 1431, 1448 (4th Cir. 1988), cert. denied, 487 U.S. 1205.

[44] See, for example, *U.S. v. Rodriguez*, 588 F.2d 1003 (5th Cir. 1979).

[45] 71 Cal. App. 3d 1, 139 Cal. Rptr. 275 (1977), cert. denied, *Manson v. California*, 435 U.S. 953 (1978).

The American tradition of trial by jury, considered in connection with either criminal or civil proceedings, necessarily contemplates an impartial jury drawn from a cross-section of the community.[46] This does not mean, of course, that every jury must contain representatives of all the economic, social, religious, racial, political and geographical groups of the community; frequently such complete representation would be impossible. But it does mean that prospective jurors shall be selected by court officials without systematic and intentional exclusion of these groups.[47]

Preventive Statistics: Using a Survey

Public opinion can be used in support of a motion for change of venue.

In Illinois, a defendant proffered a poll taken at a local shopping mall showing that 53% of the respondents thought the defendant was guilty. The appeals court ruled that the trial court erred in denying a change of venue; "such percentages illustrate the pervasive effect of saturation news coverage."[48]

On the other hand, when H.R. Haldeman and several other presidential aides were indicted for their roles in covering up the involvement of Nixon campaign officials in the burglary of Democratic National Headquarters (Watergate), their application for a change of venue was turned down by the courts. The district court of the District of Columbia ruled that a recorded comprehensive *voir dire* examination conducted by the judge in the presence of all parties and their counsel is more determinative of whether a fair and impartial jury could be impaneled than a poll taken in private and paid for by one side.[49]

---

[46] *Smith v. Texas*, 311 U.S. 128, 130 (1941); *Glasser v. U.S.*, 315 U.S. 60, 85 (1942).

[47] See also *U.S. v. Test*, 550 F.2d 577 (10th Cir. 1976); *U.S. v. Lewis*, 472 F.2d 252 (3rd Cir. 1973); *People v. Tevino*, 39 Cal.3d 667, 225 Cal. Rptr. 652 (1985).

[48] *People v. Taylor*, 447 N.E.2d 519 (1983).

[49] See *U.S. v. Haldeman*, 559 F.2d 31, 64, n. 43, (D.C. Cir. 1976), cert denied, 431 U.S. 933 (1977).

2.3.4 Standing

Parties in a civil lawsuit have standing to raise the issue of denial of a juror's rights if a party is a member of the cognizable group in question, according to *Powers v. Ohio*.[50]

In a criminal case, the defendant has standing whether or not he is a member of the cognizable group in question.[51] "Whatever his race," Justice Marshall writes for the majority in *Peters v. Kiff*, "a criminal defendant has standing to challenge the system to select his grand or petit jury on the grounds that it arbitrarily excludes from service the members of any race and thereby denies him due process."[52]

The rationale for allowing litigants to intervene on behalf of jurors is detailed in *U.S. v. De Gross*.[53]

2.4 Random Selection

What constitutes random selection? The California Trial Jury Selection and Management Act[54] states that:

> It is the policy of the State of California that all persons selected for jury service shall be selected *at random* from the population of the area served by the court; that all qualified persons have an equal opportunity, in accordance with this chapter, to be considered for jury service in the state and an obligation to serve as jurors when summoned for that purpose; and that it is the responsibility of jury commissioners to manage all jury systems in an efficient, equitable, and cost-effective manner in accordance with this chapter.

Anything can happen and often does when samples are selected at random. Red appears ten times in ten consecutive spins of the roulette wheel. Impossible? Not at all. A result like this (or one equally improbable, such as spinning black ten times in a row) can be expected once in every 500 turns of the wheel.

---

[50] 499 U.S. 400 (1991).

[51] 407 U.S. 493 (1972).

[52] Ibid. p. 504; see also *People v. Estrada* (permitting a 36-year-old defendant to raise the issue of the lack of youths on the jury).

[53] 960 F.2d 1433,1436 (9th Cir. 1992).

[54] Title 3, C.C.P. Section 191.

Justice Blackmun notes in *Ballew v. Georgia*[55] that if a minority group comprises 10% or less of a population, a jury of 12 persons selected at random from that population will fail to contain members of that minority at least 28% of the time.[56] Thus, extremes, such as a jury of 12 white males, or the ten blacks, one white, and one Hispanic who served on the O.J. Simpson criminal jury, do not constitute a legal inequity.

It is possible to avoid such extremes and create a greater likelihood that the composition of individual juries will be more representative of the community at large. In the next chapter, we consider several sampling methods that come closer to a true cross-section of the community than the method of simple random selection. To date none has been considered in the context of juror selection.

2.4.1 Errors in Sampling Methodology

How severe does an error in sampling methodology have to be for the courts to find it a basis for reversal? The Fifth Circuit dismissed indictments handed down by a grand jury finding of a substantial failure in the mechanism by which the grand jury was selected.[57] This mechanism, then experimental, but common today now that most courts are computerized, consists of three steps:

1. A master jury panel is selected in accordance with statute.
2. The number of the first juror, the seed, is selected.
3. A quasi-random computer number generator is used to select all other jurors.[58]

The first and third steps were not at issue in that case. At step two, the court clerk selected a "convenient" number. Inspection of past grand juries revealed the clerk had used the same starting seed repeatedly. Due to the nature of computerized quasi-random number generators, the numbers of the other "randomly selected" jurors had been repeated as well. Although the sequence looks random in character and passes all statistical tests for randomness, it is actually fixed. To ensure true randomness, the starting point in the sequence must be selected at random by some chance device. Today, most jury selection software chooses a starting random number by accessing the computer's internal clock; the software

---

[55] 435 U.S. 223, 236–237 (1978).
[56] We go through the calculations in Chapter 5.
[57] *U.S. v. Northside Realty Assoc.*, 659 F.2d 590 (5th Cir. 1981).
[58] For details, see Boswell et al., 1993 and VanDyke, 1977.

takes the time to the nearest millisecond, divides by a very large number and uses the remainder as the starting point.[59]

Not all deviations from random selection result in reversal. In *People v. Viscotti*,[60] the issue was whether the trial court erred in taking the first 12 jurors from the panel rather than selecting 12 at random. The court held that a material departure from statutory procedures had not occurred as the panel had been selected at random from the population. This decision is mathematically correct. Once the cake batter is mixed, further mixing is redundant and unnecessary.[61]

2.5 Summary

Parties in a civil law suit and the defendant in criminal court have standing to raise the issue of denial of a juror's rights. To establish a *prima facie* violation, three points must be demonstrated:

1. The group alleged to be excluded is a distinctive cognizable group — examples include race, gender, religion, political party, or combination thereof.
2. Representation in this group in a series of venires from which juries were selected is not fair and reasonable by comparison with the population in the local judicial district.
3. The underrepresentation is due to systematic exclusion at some point in the juror-selection process.

The composition of any specific panel (sample) is irrelevant, and challenges must be based on past patterns of discrimination. With appropriate changes in wording, these points apply to samples of any type taken for purposes other than jury selection — for example, to establish trademark infringement or discrimination in hiring.

2.6 To Learn More

For more discussion on jury selection, see Beale [1984], DiPrima [1995], Harrison [1990], Starr and Jordan [1993], and Zeisel and Kaye [1997]. For instructions on change-of-venue surveys, see Niestzel and Dillehay [1986; pp. 70–79].

---

[59] This is the same method used by the computer to decide what hand of solitaire or FreeCell you'll play next.

[60] 2 Cal. 4th 1, 5 Cal. Rptr. 2d 495 (1992).

[61] See, for example, Good [2000] or Finklestein [1973].

Sample Selection

Statisticians would say sample selection is fair and reasonable if:

■ Each individual in a population has an equal probability of being selected.
■ Selections are independent of one another.
■ Selections are made in random order.
■ An identical method of measurement is used on each item if measurements are involved.

Chapter 3

Sample and Survey Methodology

3.1 Concepts

In this chapter, you will learn methods for reducing the cost of samples and surveys through cluster and stratified sampling and for improving their accuracy through dual-system estimation and subsampling of nonrespondents. You will also learn how to detect and forestall sample bias and to determine circumstances under which missing data is acceptable in a courtroom.

3.2 Sampling Methodology

In Chapter 2, we saw that simple random sampling, while fair in the aggregate, could result in the selection of a specific jury panel that is not representative of the population as a whole. For example, the jury that served in O.J. Simpson's criminal trial consisted of ten African-Americans, one Caucasian, and one Hispanic, in contrast to the far more diversified population of Los Angeles from which it was chosen. As an alternative method of selection, Fukurai et al. [1991] propose the use of *stratified sampling*. Here is how this would work.

Suppose the census shows that six of 12 residents of an area are Caucasian, four are Hispanic, and two are African-American. Under the Fukari proposal, each jury would consist of six residents selected from the Caucasian pool, four from the Hispanic, and two from the African-American.

Stratified sampling should be employed whenever the composition of the population to be sampled varies widely from place to place or time to time within the period of the study. For example, in recoupment cases, stratified sampling would be mandated statistically if the sample covered varied programs in varied locations (different procedures, places of service, or performing providers) or extended over a period during which either the provider or the government changed its practices or policies.[1]

In *Ratanasen*,[2] the defendant physician contended that simple random sampling, while an appropriate way to create a sample from a homogenous population, is not appropriate in a situation such as his involving a heterogeneous population, where stratified random sampling should have been used.

In support of this proposition, he cited *Grier v. Kizer.*[3]

> An illustration given is of a sample drawn from a box containing 500 pounds of oranges and diamonds. A random ten-pound sample might yield a misleading value if it contained a dispro-portionate number of diamonds. Stratified sampling, as contrasted with random sampling, would draw separate samples of diamonds and oranges.[4]

During an evidentiary hearing on this objection, the court heard from a research program specialist for the state who testified that the simple random sampling method chosen for this case was "appropriate, valid, and reliable." The physician's own expert testified that the appropriate method to sample would have been stratified random sampling, and that the state's method was invalid and unreliable. The bankruptcy judge ruled that simple random sampling was a valid method to use.

HCA Health Services v. Kansas[5] had a different result. Again, the court reviewed an order of recoupment based on a simple random sample. HCA Health Services maintained that a stratified sample should have been used because of differences in the types of procedures, providers of service, and places of treatment. The administrative hearing officer and later the trial court took the position that, absent a second audit based on a stratified sample, HCA Health Services had failed to show that the Kansas agency's survey methods were unreliable.

---

[1] See also Section 13.3.2.

[2] See discussion in Section 1.1.

[3] 219 Cal. App. 3d 422, 268 Cal. Rptr. 244 (1990).

[4] Id. at 268 Cal. Rptr. at 247, n. 3. The court declined to review the statistical validity of the method, reversing on other grounds, at 255.

[5] 900 P.2d 838 (Kan. Ct. App. 1994).

The appellate court found the agency had failed to comply with its own policy directives that set forth differing procedures for homogeneous and heterogeneous populations. "According to the policy, differing procedure codes, places of service and performing providers are characteristics indicating heterogeneity."[6] Although the agency had a policy for the use of stratified sampling, it never developed a procedure for its use.[7]

3.2.1 Cluster Sampling

Whatever the merits of stratified sampling for jury selection, it has proven its value many times over in survey work, particularly when coupled with a second sampling technique known as *clustering*. For example, in a trademark dispute, you might want to survey the public to establish that "8-Up," a competitor's new citrus-flavored drink, is too easily confused with your client's long-established brand. Or you might want to assemble a series of mock jury panels to assess various trial strategies. You would begin the survey process by dividing the selected geographical area (the applicable country, state, or county) into *strata* such as urban poor, rural poor, middle class, ethnic groups, and so forth. If your survey necessitates going door to door, you might want to further subdivide each stratum into *clusters* based on geography. For example, you might designate certain zip codes as urban poor and others as African-American middle class. Next, you would select a zip code at random from each stratum, and from within that zip code, certain households at random to interview. This cluster approach dramatically reduces travel and personnel expenses; it is statistically valid, provided both clusters and individuals within clusters are selected at random. Because you are selecting separately from each stratum, the resultant sample will be representative in theory and in fact.

As an example, to buttress a dispute over brand identification, Zippo Manufacturing used cluster sampling to reduce the cost of an extremely large survey of the lighter-buying public. Selecting 53 demographically distinct localities across the U.S. as the basis of their survey, they chose 100 clusters of about 150 to 200 dwelling units within each locality, and then selected both clusters and individuals within clusters at random to make up a total of 500 respondents. "The weight to be given a survey, assuming it is admissible, depends on the procedures by which the survey was created and conducted."[8] Although the court expressed its approval

---

[6] Id. at 849.

[7] Id.

[8] *Zippo Manufacturing v. Rogers Imports*, 216 F. Supp. 670 (S.D. N.Y. 1963); see also *Rhodes Pharmacal Co. v. Federal Trade Comm.*, 208 F.2d 382, 387 (7th Cir. 1953), rev'd in part on other grounds, 348 U.S. 940 (1955).

of the design and implementation of the surveys, it ruled the evidence so derived failed to establish Zippo Manufacturing's claims.

The purpose of cluster sampling is to save money, not to increase accuracy. Cluster sampling alone without the use of predesignated strata is like a blind man considering an elephant. Perhaps he'll think it's a snake if he takes hold of the elephant's tail, or perhaps a tree trunk if he puts his arms around the elephant's leg. Cluster sampling without the use of strata is essentially what you get in so-called man-in-the-street interviews, or that infamous survey in the 1948 presidential election limited to those in the phone book that led to declaring the losing candidate, New York Governor Thomas Dewey, the winner over ultimately successful Harry Truman.

3.2.2 The Fight over the Census

> Los Angeles, March 3 1998 (UPI). Los Angeles' top prosecutor says he will fight House Speaker Newt Gingrich's lawsuit seeking to ban the use of statistical sampling in the census in the year 2000.

After the 1990 census was complete, big city mayors complained that the numbers of urban poor, particularly the homeless, had been grossly underestimated, thus depriving urban areas of the funds and representation to which a true census would have entitled them. As mandated by statute, census takers had gone from house to house to interview or to pick up previously completed survey forms. There were those who would not respond to such a census, the critics argued, whether through fear of the law or antipathy to authority. Others, living in boxes or under freeway underpasses, could not be found. A recount using the same methods would have met with the same obstacles.

At issue in *City of New York v. Dept. of Commerce*[9] was whether the percentage of respondents could be increased through subsampling and statistical means to improve census reliability. Many experts in survey methods testified on both sides.[10] In the face of such conflict among the experts, the court ruled that the decision should be left to the Secretary of Commerce.[11]

---

[9] 822 F. Supp. 906 (E.D. N.Y., 1993).

[10] The arguments of four of these statistical experts may be found in Volume 34 of *Jurimetrics*, 1993, 64–115.

[11] Ibid. at 931.

The proposed subsampling procedure is called *dual-system estimation* and is the standard method by which naturalists estimate the sizes of wildlife populations.

Suppose we count and tag 200 fish in a lake. We know we have not counted all the fish — many are bound to be hidden in the reeds. We take a second random sample of 150 fish and find that 125 of them, or ⁵/₆, bear tags. We reason that the fish in the lake are distributed in the same ratio of 5:1, 200 that we counted the first time, and 40 that were not counted, for a total of 240 fish.

New York City wanted the Census Bureau to undertake a similar post-census study; instead of drawing its sample from the entire population of the U.S. — a ruinously expensive proposition — the stratified cluster method would be used. S.E. Fienberg [1993], a statistician testifying on behalf of New York City, admitted that the post-census study relied on three crucial assumptions:

1. Perfect matching — the ability to establish with certainty whether someone had already been "tagged," that is, included in the original census;
2. Independence of the post-census study and the census; and
3. Homogeneity — the probability of including an individual in the survey who escaped being counted the first time would be the same as the probability of surveying an individual who had been included in the original census.

We can immediately add a fourth assumption seized on by the Secretary of Commerce [Freedman, 1994]:

4. The cluster we choose to sample from — a farming community in Iowa or a block of condemned tenements in Newark — must be representative of the strata as a whole.

Statisticians testifying for the defense immediately seized on the weaknesses of these assumptions:

■ Clerical and other processing errors would interfere with perfect matching.
■ People are not fish; those missed by the original census would be just as likely to be missed on the second go-around.
■ The chosen strata were not representative. An urban ghetto in New York City is not the same as an urban ghetto in Santa Fe, New Mexico.
■ Additional sampling error due to the luck of the draw would make the post-census survey much less reliable than the original census.

These are all questions of fact,[12] and the trial judge ruled that the Secretary of Commerce was the appropriate individual to decide among the competing claims.

3.3 Increasing Sample Reliability

Assume you selected your samples and sampling units at random, and ensured (by the way you sampled) that the responses are independent of one another. Is this good enough? Not quite. The results of the survey may depend on *how* you ask the questions and whether you record the answers correctly.

3.3.1 Designing the Questionnaire

Macmillan, Inc. was charged by Federal Trade Commission (FTC) investigators with running a correspondence school that promised far more than it could deliver.[13] The FTC supported its contention with a survey of former students made by the Resource Planning Commission (RPC).

The survey had been taken in three stages. First, questionnaires were mailed to former students. After a lapse of time, a second set of questionnaires was mailed to those who failed to respond the first time. Finally, a telephone survey was conducted among those who failed to respond to either mailing. A statistical test revealed no significant differences between those who responded immediately and those who had to be pressured to respond. Nonetheless, the appeals board rejected the use of the survey on three grounds:

1. A cover letter accompanying the mailing served to prejudice respondents against the school. (The cover letter read like a veritable "call to arms" for dissatisfied students. Printed on FTC letterhead, it began, "The Bureau of Consumer Protection is gathering information from those who enrolled in ... to determine if any action is warranted." This letter was signed by an FTC attorney.)
2. The multiple-choice questions were poorly designed and offered inadequate choices. (Fortunately, many books and journal articles on this topic are available to help the attorney today. See, for example, Dutka, 1982 and Fink and Kosecoff, 1998.)
3. Respondents were not permitted (as they would have been on the witness stand) to say "I don't know" or "I can't recall." (Two to

---

[12] See Section 1.1.
[13] See *Macmillan, Inc.*, 96 F.T.C. 208 (1980).

five years elapsed between the time some respondents were enrolled in the school and the date by which they were asked to respond to the survey — a sufficiently lengthy interval that many individuals honestly would not remember.)

3.3.2 Data Integrity

Virtually every statistical procedure relies on the individual observations being independent of one another.[14] In *Toys "R" Us, Inc. v. Canarsie Kiddie Shop, Inc.*,[15] the court rejected the results of a survey in which some of the interviews were conducted in a bowling alley, thus allowing some of those waiting to be interviewed to overhear the substance of the interview.

Courts are not tolerant of errors in data entry and encoding.

> Many coding errors ... affected the results of the survey.[16]

> [E]rrors in EEOC's mechanical coding of information from applications in its hired and nonhired samples also make EEOC's statistical analysis based on this data less reliable .... [The EEOC] consistently coded prior experience in such a way that less experienced women are considered to have the same experience as more experienced men [and] has made so many general coding errors that its data base does not fairly reflect the characteristics of applicants for commission sales positions at Sears.[17]

> Although plaintiffs show that there were some mistakes in coding, plaintiffs still fail to demonstrate that these errors were so generalized and so pervasive that the entire study is invalid.[18]

Computers crash and typing errors occur. The good news is that advances in computer technology and a decline in the cost of mass storage have largely eliminated the problems inherent in coding and decoding. No longer is there a need for cryptic abbreviations: Does *S* stand for "satisfactory"?

---

[14] See Section 5.4 for a formal definition of independence.

[15] 559 F. Supp. 1189, 1204 (E.D. N.Y. 1983).

[16] *G. Heileman Brewing Co. v. Anheuser-Busch, Inc.*, 676 F. Supp. 1436, 1486 (E.D. Wis. 1987).

[17] *EEOC v. Sears, Roebuck & Co.*, 628 F. Supp. 1264, 1304-1305 (N.D. Ill. 1986), aff'd, 839 F.2d 302 (7th Cir. 1988).

[18] *Dalley v. Michigan Blue Cross-Blue Shield, Inc.*, 612 F. Supp. 1444, 1456 (E.D. Mich. 1985).

Or "sick"? And there never was an excuse for coding male and female as *1* and *2* when *M* and *F* were far less open to error and ambiguity.

The most effective way to eliminate errors in data entry is to do so at the time of entry, using readily available and inexpensive computer software that checks and validates the data as it is entered. Ambiguities can be resolved while the cognizant personnel are still readily available.

3.4   How Much to Tell the Court

Must all details of a survey be completely revealed? *Florida Bar v. Went For It, Inc.*[19] concerned the rights of attorneys to send targeted direct-mail solicitations to victims and their relatives within 30 days following an accident or disaster, regardless of rules to the contrary imposed by the Florida Bar. The Florida Bar introduced into evidence a summary of a survey of irate consumers, and the defense objected to the use of a summary in place of the original survey.

Justice O'Connor wrote on behalf of the majority, "We do not read our case law to require that empirical data come to us accompanied by a surfeit of background information." As statisticians anxious to have our painstaking efforts appreciated, we hope the reasoned dissent by Justice Kennedy, in which Justice Stevens, Justice Souter, and Justice Ginsburg joined, will prevail:

> The burden of demonstrating the reality of the asserted harm rests on the State. Slight evidence in this regard does not mean there is sufficient evidence to support the claims. Here, what the State has offered falls well short of demonstrating that the harms it is trying to redress are real, let alone that the regulation directly and materially advances the State's interests. The parties and the Court have used the term "Summary of Record" to describe a document prepared by the Florida Bar, one of the adverse parties, and submitted to the District Court in this case. This document includes no actual surveys, few indications of sample size or selection procedures, no explanations of method-ology, and no discussion of excluded results. There is no description of the statistical universe or scientific framework that permits any productive use of the information the so-called Summary of Record contains.[20]

---

[19] 515 U.S. 618 (1995).
[20] Id. at 640.

3.5   Missing Data and Nonresponders

The FTC forestalled at least some of the objections to its survey in *Macmillan* by deliberately polling those who failed to respond the first time, then showing there were no statistically significant differences between those who responded immediately and those who had to be pressured to respond. But what if you can't reexamine nonresponders? What if the data is simply missing?

The Equal Employment Opportunity Commission (EEOC) alleged that Eagle Iron Works assigned African-Americans to unpleasant work tasks because of their race and discharged African-Americans in greater numbers than Caucasians, again because of their race. The EEOC was able to identify only 1200 of 2000 past and present employees by race although the races of all 250 current employees could be identified. The court rejected the contention that the 250 current employees were a representative sample of all 2000; it also rejected the EEOC's unsubstantiated contention that all unidentified former workers were Caucasian. "The lack of a satisfactory basis for such an opinion and the obvious willingness of the [expert] witness to attribute more authenticity to the statistics than they possessed, cast doubts upon the value of [his] opinions."[21]

Similarly, the plaintiff's survey was rejected in *Bristol Meyers v. FTC*,[22] due to failure to follow up the 80% who did not respond.

The enormous amount of data involved in a third discrimination suit, *Vuyanich v. Republic National Bank*,[23] involving hundreds of variables and thousands of employees, could never have been analyzed without a computer. As might have been expected in a study of that magnitude, many of the questionnaires were only partially completed, contained crossed-out items or unreadable corrections, or simply were not returned. The defendant argued that employees who did not complete the questionnaires were different from those who did. These employees may not have understood what they were asked to do or may not have wanted to cooperate in the bank's defense. The court said that high completion rates (99.7% partially completed and 93.7% fully completed) argued to the contrary.

> The challenging party bears the burden of showing that errors or omissions bias the data, i.e., that erroneous or omitted items are not distributed in the same ways as items which are present and correct.[24]

---

[21] *Eagle Iron Works*, 424 F. Supp. 240, 246-247 (S.D. Ia. 1946).

[22] 185 F.2d. 258 (4th Cir. 1950).

[23] 505 F. Supp. 224, 255-258 (N.D. Tex. 1980).

[24] Id. at 255.

The defendant demonstrated an omission rate of 31% and an error rate of 10% for at least some of the items in the database, but failed in the court's opinion to show that these omissions or errors were not randomly distributed among sexes or races, or that over- or under-estimates resulted.

3.6 Summary

Acceptable sampling methods include stratified cluster and simple random sampling.

1. The sampling method must be appropriate; testimony from a statistician is essential in making this determination and should make clear both the principles and the procedures by which the survey was conceived and conducted.[25]
2. Conduct of the survey should be independent of the attorneys involved in the litigation.[26]
3. Questions must be shown to be free of bias; testimony from a specialist in survey design and interpretation would be helpful in making this determination.
4. Errors in data collection, data entry, and data storage should be minimal. Procedures for error detection and correction should be implemented.
5. The sample size must be adequate; see Chapter 9.
6. Selection procedures and statistical methodology must be well documented; excluded results must be accounted for. The challenging party bears the burden of showing that errors or omissions bias the data.

---

[25] See *Zippo Manufacturing v. Rogers Imports*, 216 F. Supp. 670, 681 (S.D. N.Y. 1963).
[26] *Pittsburgh Press Club v. U.S.*, 579 F.2d 751 (3rd Cir. 1978).

Preventive Statistics

Surveys raise as many questions as they resolve and are open to attack from all sides. Was the sampled population germane? Was the sampling methodology appropriate? Are the survey questions free from bias?[27] Even expertly designed studies will be criticized. Adhere to the guidelines in the *Federal Rules of Evidence*[28] and the *Federal Judicial Center's Manual for Complex Litigation.*[29] Becker [1991] proposes the following guidelines:

■ A survey should be designed, executed, and analyzed through significant input from an expert in the field of polling.
■ Samples should be selected from the appropriate universe.
■ The number of respondents should be statistically significant.
■ Questions should be appropriately drafted and arranged so as to avoid bias or lead respondents.
■ Questions should be pretested. (During discovery, determine how your opponent's surveys have been pretested and modified.[30])
■ Trained persons should conduct interviews.
■ Respondents should be unaware of the survey's purpose.
■ Data collection and processing should include safeguards to minimize errors in transcription.
■ The conclusions drawn should be supported by the poll results.

---

[27] See, for example, Cochran [1977], Converse and Presser [1986], Dutka [1982], and Fink and Kosecoff [1998].
[28] 803(24); see also 703, Advisory Committee Note.
[29] §2.712 at 118 [1982].
[30] See, for example, Schroeder [1987].

Chapter 4

Presenting Your Case

4.1 Concepts

This chapter introduces the basic concepts of descriptive statistics, the center or average of a sample and its relation to the population average, the precision of sample estimates, and changes in rates.

4.2 The Center or Average

Now that your sample has been introduced into evidence, what are you going to tell the judge and jury about it? A good first step would be to introduce a chart or graph. A picture is worth a thousand words (see, for example, Cleveland [1985, 1993], Good [2000; Chapter 1], and Whittaker [1990]).

When making comparisons, focus the listener's attention on a single value. Most often that value will be some kind of average or central value.

The easiest of the averages to compute is the *median*, that value which is at the midpoint or 50th percentile of the sample; 50% of the observations will be larger than the median and 50% will be smaller. For example, the median of the sample consisting of the three observations 1, 2, and 6 is 2.

Most readers will already be familiar with the *arithmetic mean*, the sum of the observations divided by their number. The mean represents an equilibrium or pivot point in that the sum of the deviations of the individual observations from the mean is zero. The arithmetic mean of the 1, 2, and 6 sample is $(1 + 2 + 6)/3 = 9/3 = 3$.

Some judges feel the median is more representative than the mean, particularly when the intent is to measure income or cost. This is because one or two very large or very small observations can affect the value of the mean, while leaving the median relatively unaffected.[1] Still, the arithmetic mean remains the best choice when extrapolating from the sample to the population total. For example, if the arithmetic mean of the out-of-pocket costs of plaintiffs in a class action suit is $1253 and the suit has 1000 potential plaintiffs, then 1000 × $1253 should be set aside to cover plaintiffs' costs.

4.2.1 *Extrapolating from the Mean*

In *U.S. v. Shonubi*,[2] Charles Shonubi had a total of 427.4 grams of heroin sealed inside condoms in his digestive tract when he was arrested at Kennedy Airport in 1991.

> A forensic chemist had selected at random four of the 103 balloons passed by Shonubi after his arrest, determined the [arithmetic mean] average weight of the heroin contained in these four balloons, and then multiplied that average [mean] by 103 to conclude that the weight of the heroin contained in all 103 balloons was 427.4 grams. This approach rested on the assumption that the four balloons selected were representative of the entire 103 balloons, an assumption that, in turn, rested on subsidiary assumptions that [i] each of the 103 balloons contained heroin, [ii] the average quantity of heroin in each of the four balloons selected was the same as the average quantity in all of the 103 balloons, and [iii] the average purity of the heroin in the four balloons selected was the same as the purity of the heroin in all of the 103 balloons because all the heroin came from the same batch of heroin.[3]

4.2.2 *The Geometric Mean*

The *geometric mean* is appropriate in two sets of circumstances: (1) when losses or gains can best be expressed as a percentage rather than a fixed

---

[1] See, for example, *CSX Transport. Inc. v. Board of Public Works of West Virginia*, 95 F.3d 318 (4th Cir. 1996); *Clinchfield R.R. Company v. Lynch*, 527 F. Supp. 784 (E.D. N.C. 1981), aff'd, 700 F.2d 126 (4th Cir. 1983).
[2] 103 F.3d 1085 (2nd Cir. 1997). See also Section 6.4.
[3] Ibid. at 1091.

value, and (2) when rapid growth is involved as in the growth of a bacterial or viral population.

There are many different ways to compute a geometric mean. The simplest can be gleaned from the following example. A broken sewer main closes all businesses on a busy street for two days. Individual businesses lose from $400 to $4000. The geometric mean loss can be obtained by taking the arithmetic mean of the individual losses expressed in percentage terms, then converting this percentage of the total loss to dollars. But why bother? If individual losses ranged from 18 to 23% of mean daily revenue with an arithmetic mean of 20%, then the proper class action award would be 20% of mean daily revenue, rather than any single fixed dollar amount.

Because bacterial populations can double in number in only a few hours, many government health regulations utilize the geometric mean rather than the arithmetic mean.[4] A number of other government regulations also use it although the sample median is far more appropriate.[5] In contexts where the changes are proportional rather than additive, the geometric mean is computed by taking the logarithms of the individual observations and the arithmetic mean of the logarithms, then computing the antilog of the mean. An example of such a calculation using an Excel spreadsheet is given in Table 4.1. The original observations are in Column A. Their logarithms are in Column B. The formula in Row 6 Column C is for the geometric mean.

Table 4.1 An Excel Spread Sheet

	A	B	C
1	100.00	4.605	$= \ln(A_1)$
2	110.00	4.700	$= \ln(A_2)$
3	123.00	4.812	$= \ln(A_3)$
4	98.00	4.585	$= \ln(A_4)$
5	85.00	4.443	$= \ln(A_5)$
6	102.42		$= \exp [\text{Sum}(B1:B5)/5]$

The Michigan Department of Natural Resources wanted the Federal Energy Regulatory Commission to force the Indiana–Michigan Power Company to take measures to reduce the number of fish trapped in the company's turbines and to compensate the state of Michigan for the fish

---

[4] See, for example, 40 CFR Part 131, 62 Fed. Reg. 23004 at 23008 (April 28, 1997).
[5] Examples include 62 Fed. Reg. 45966 at 45983 (concerning the length of a hospital stay) and 62 Fed. Reg. 45116 at 45120 (concerning sulfur dioxide emissions).

that were killed. A sample taken over several days selected at random was used to estimate the number of fish killed annually. Using the arithmetic mean, the state set that number at 14,866. Claiming that kills were unusually high on some of the sample days, thus biasing the arithmetic mean, the power company used the geometric mean of the sample to obtain the much lower number of 7750 killed fish. The Federal Energy Regulatory Commission's ruling in favor of the power company was upheld by the D.C. Circuit appeals court.[6]

The Center of a Population

Median: the value in the middle; the halfway point; that value which has equal numbers of larger and smaller elements around it.

Arithmetic mean or arithmetic average: the sum of all the elements divided by their number or, equivalently, that value such that the sum of the deviations of all the elements from it is zero.

Mode: the most frequent value. If a population consists of several subpopulations, there may be several modes.

4.2.3 The Mode

The *mode* or most frequent value is not recommended for estimation or summary purposes; it is useful when trying to distinguish between a homogeneous population that has only one mode and a multimodal mixture of populations such as the population depicted in Figure 4.1.

4.3 Measuring the Precision of a Sample Estimate

The court expects us to provide both an average based on our sample and some measure of the accuracy of our average. Three approaches are described in what follows.

---

[6] *Kelley v. Federal Energy Regulatory Comm'n.*, 96 F.3d 1482, 1490, fn. 7 (D.C. Cir. 1996).

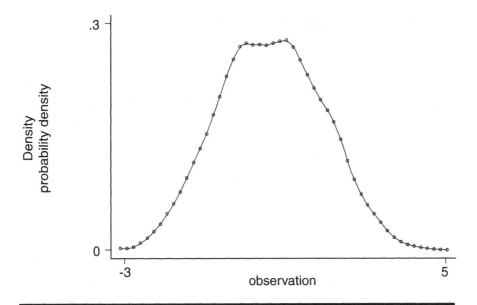

Figure 4.1 Frequency distribution of a mixture of two populations.

4.3.1 Standard Deviation

We can seldom establish the accuracy of an estimate, for example, how closely the sample median comes to the unknown population median, but we may be able to establish its precision, that is, how closely the estimates derived from successive samples resemble one another.

The most common approach is based on the *variance* of the sample. If the individual observations are denoted by X_1, X_2, and so forth up to X_n and the sample mean by

$$\bar{X} = \frac{X_1 + X_2 + \dots + X_n}{n} = \sum_{i=1}^{n} x_i/n$$

then the variance:

$$V = \sum_{i=1}^{n} (\bar{X} - X_i)(\bar{X} - X_i)/(n-1)$$

If X is in seconds, then V is in seconds squared. The *standard deviation* $\sigma = \sqrt{V}$ is in the same units as the observations.

The mean is more precise than any single observation; its standard deviation, known as the standard error of the mean, or more commonly as the *standard error*, is $\sigma/\sqrt{n}$. If the standard deviation is 4, and we take 16 observations, the standard error of their mean will be 1.

4.3.2 Bootstrap

The standard deviation is a useful measure of precision only for populations that are symmetric about their mean. For skewed populations such as incomes, property values, and fish kills, with the majority of observations at the low end, the standard deviation can be misleading. A much better approach is to *bootstrap* from the sample.

When we bootstrap, we treat the original sample as a stand-in for the population and resample from it repeatedly, with replacement, 1000 times or so, recomputing the average each time. In practice, the computer does the work of resampling. If you have no computer, you can achieve the same result by writing each of the observations on a separate slip of paper, putting all the slips into an urn, drawing one slip at a time from the urn, then replacing the slip in the urn as soon as you've recorded the number written on it. Repeat this procedure until you have a bootstrap sample of the same size as the original.

For example, here are the heights of students in a sixth-grade class, measured in centimeters and ordered from shortest to tallest: 137.0, 138.5, 140.0, 141.0, 142.0, 143.5, 145.0, 147.0, 148.5, 150.0, 153.0, 154.0, 155.0, 156.5, 157.0, 158.0, 158.5, 159.0, 160.5, 161.0, 162.0, and 167.5.

The median height lies somewhere between 153 and 154 centimeters. If we want to extend this result from a single classroom to the population of all sixth graders, we need an estimate of the precision of this average.

The first bootstrap sample, arranged in increasing order of magnitude for ease of reading, might look like this: 138.5, 138.5, 140.0, 141.0, 141.0, 143.5, 145.0, 147.0, 148.5, 150.0, 153.0, 154.0, 155.0, 156.5, 157.0, 158.5, 159.0, 159.0, 159.0, 160.5, 161.0, and 162.

Several of the values have been repeated as the technique involves sampling with replacement. The minimum value in this sample is 138.5, higher than the minimum of the original sample. The maximum at 162.0 is lower than the original, while the median remains unchanged at 153.5.

The following is a further bootstrap sample: 137.0, 138.5, 138.5, 141.0, 141.0, 142.0, 143.5, 145.0, 145.0, 147.0, 148.5, 148.5, 150.0, 150.0, 153.0, 155.0, 158.0, 158.5, 160.5, 160.5, 161.0, and 167.5. In this second sample, we again find repeated values. The minimum, maximum, and median are 137.0, 167.5, and 148.5, respectively.

The medians of 50 bootstrapped samples drawn from our sample of sixth graders ranged between 142.25 and 158.25 with a median of 152.75.

|
142.25   Medians of Bootstrap Samples   158.25

Figure 4.2   One-way scatterplot of 50 bootstrap medians derived from a sample of heights of 22 students in Dr. Good's sixth-grade class.

They provide a feel for what might have been the results had we sampled repeatedly from the original population and are sufficient to convey to the court the precision of our estimates.

4.3.3   Coefficient of Variation

The problem with the variance is that if the observations are in feet or dollars or seconds, then the variance V is in square feet, square dollars, or square seconds — whatever those are. The standard deviation is the square root of the variance and is measured in the same units as the observations. By expressing the standard deviation as a percentage of the mean, or equivalently, by computing the *coefficient of variation*

$$ C = \frac{\sqrt{V}}{\overline{X}} , $$

we obtain a unit-free number that can be used to compare the accuracy of samples collected at different times or by different means. It can also be used for quality control purposes and for classification.

Automobile tires come in a variety of widths and heights. Rather than set individual quality control standards for each combination, the National Highway Traffic Act specifies that the coefficient of variation cannot exceed a certain value. As we shall see in Section 10.2, requiring that this coefficient be below 0.005 is equivalent to specifying that 95% of the observed values be within 1% of the mean. This provision of the act was upheld on appeal in the Sixth Circuit.[7]

Just as fingerprints can be used to distinguish individuals, the combination of the mean and the coefficient of variation can be used to distinguish certain populations. The Federal Trade Commission charged Forte-Fairbairn, Inc. with mislabeling and selling baby alpaca fibers as "baby llama" fibers. Forte-Fairbairn was able to demonstrate in court that while the coefficient of variation of the width of baby alpaca fibers ranged from 18 to 25%, those of baby llama fibers and the company's own product ranged from 28 to 35%. The complaint was dismissed.[8]

---

[7] See, for example, *B.F. Goodrich Co. v. U.S. Dept. of Transportation*, 541 F.2d 1178, 1189 (6th Cir. 1976).

[8] *In Re Forte-Fairbairn, Inc.*, 62 F.T.C. 1146 (1963).

The Precision of a Sample

Range: the minimum and maximum values in a sample.

Standard deviation: measured in the same units as the observations; the square root of the mean of the squared deviations about the sample mean.

Standard error: the standard deviation of the sample mean is the standard deviation of an individual observation divided by the square root of the number of observations.

Coefficient of variation: a unit-free value; the sample standard deviation divided by the sample mean.

We return to the topic of precision in Chapter 9 when we discuss sample size.

4.4 Changes in Rates

Rates and proportions will be discussed further in the chapters that follow.

Twenty-five percent of the Caucasians in a community are called for jury duty at one time or another, but only 5% of the Hispanics are called. Ninety percent of males are promoted in the first two years after hiring, but only 10% of females are promoted. The rate of childhood leukemia among the general population is a little more than one case among 100,000 individuals; but in Woburn, MA, the setting for the book and the movie titled *A Civil Case,* the incidence of leukemia was more than 20 cases per 100,000 individuals.

How are we to make a comparison between two such rates? At issue in *Craig v. Boren*[9] was the right of college-age males to drink in Oklahoma. College-age females already had that privilege and the inequity seemed an obvious case of discrimination. The state of Oklahoma argued that the law only reflected a clear difference in the behaviors of young men and women. About 2% (427) of the total number of males of college age were arrested in 1973 for driving under the influence, whereas only 24 females, about 0.18% of the relevant population, were arrested for the same offense.

To provide meaningful comparisons, statisticians often recommend using the odds ratio, the passing rate divided by the failure rate, 98%

---

[9] 429 U.S. 1124 (1976).

divided by 2% = 48 for men and 99.82% divided by 0.18% = 554 for women. The odds ratio avoids the trap fallen into by the court in *Davis v. City of Dallas*[10] when it observed that a "7% difference between 97% and 90% ought not to be treated the same as a 7% difference between, e.g. 14% and 7%, since the latter figure indicates a much greater degree of disparity." This is not the case because pass rates of 97% and 90% immediately imply failure rates of 3% and 10%.

The court in *Craig v. Boren* ruled that, "While such a disparity is not trivial in a statistical sense, it hardly can form the basis for employment of a gender line as a classifying device."

4.4.1 Comparative versus Absolute Disparity

In *People v. Harris*,[11] defendants disputed the fairness of the manner in which the jury was selected by comparing the racial composition of the population at large with the composition of the jury panel. California courts supported the defense's use of comparative rather than absolute disparity, citing Karys [1977]. As an example of the distinction, if a population consists of 12% African-Americans, and a series of panels contains only 4%, the absolute disparity is 8%, but the comparative disparity is 66%.

4.5 Summary

In presenting your case, begin with a picture, then focus the court's attention on one or two numbers — a mean, a median, a range, or an odds ratio. The median is recommended for descriptive purposes, but arithmetic or geometric means should be used if additional calculations are required. The geometric mean should be used when dealing with proportional rather than additive changes.

Be prepared to provide sample sizes and measures of precision. The standard error is recommended for use with symmetric distributions; with samples of 20 or more, bootstrapping will yield a more vivid and accurate picture. Only after laying the foundation through the use of these measures should you go on to discuss more sophisticated relationships such as those outlined in Chapters 11 and 12.

For other informative examples of the application of descriptive statistics in the law, see Barnes [1983].

---

[10] 487 F. Supp. 389 (N.D. Tex. 1980).

[11] 36 Cal. 3d 36 (1984), 201 Cal. Rptr. 782, cert.denied, 469 U.S. 965, appeal to remand, 191 Cal. App. 3d 819, 236 Cal. Rptr. 680, appeal after remand, 217 Cal. App. 3d 1332, 236 Cal. Rptr. 563.

PROBABILITY

II

Statistical evidence has no magic properties.

James C. Hall, Judge, 9th Circuit[1]

Statistics are not irrefutable; they come in infinite variety and, like any other kind of evidence, they may be rebutted. In short, their usefulness depends on all the surrounding facts and circumstances.

Justice Potter Stewart, U.S. Supreme Court[2]

Part I illustrated how a series of jurists and legislatures, working in harmony with and responding to input from statisticians, developed sample selection criteria that have become guidelines for statisticians.

In Part II, devoted to the theory and legal application of probability, the situation is quite different. Universal principles, the so-called laws of probability defined in Chapter 5, are given different weights, depending on the forum. At least four distinct situations arise:

- The criminal courts, discussed in Chapter 6
- Civil law with the exception of toxic torts, discussed in Chapter 7
- Toxic torts, discussed in Chapter 8
- Administrative hearings on discrimination where (as discussed in Chapter 12) probability and statistics, although necessary, are not sufficient.

---

[1] *Williams et al. v. General Motors Corp.* 656 F.2d 120 (1981) at 130, n 14.
[2] *Teamsters v. U.S.*, 431 U.S. 324 at 340 (1977).

Chapter 5

Probability Concepts

This chapter introduces essential probability concepts and provides formal definitions for equally likely and mutually exclusive events, conditional and unconditional probabilities, independence, the product rule, and sampling with and without replacement.

5.1 Equally Likely, Equally Frequent

Consider a small six-sided cube made of a single homogeneous material, ivory perhaps, or, more likely, plastic. If you have gone to Las Vegas or Atlantic City, or participated in a game of backgammon, craps, Chutes and Ladders, or Monopoly, then you know the cube is a die. Since the six sides of the die are equivalent in every respect (except for the numbers engraved upon them), we can assume they are equally probable; that is, if we roll the die over and over again, each of the six sides will turn up with equal frequency. Thus, the probability of rolling a 5 is one in six.[3] The same principle applies to every other chance device. What is the probability of drawing the ace of spades from a well-shuffled deck of 52 cards? One in 52.

One hundred thousand people live in my court district. If jury members are selected at random, what is the probability that I will be the next person summoned for jury duty? One in 100,000. (Any other result, as we saw in Chapter 2, holds the potential for a verdict reversal.)

---

[3] We assume that some unscrupulous person has not weighted the die so that one side is more likely to appear.

5.2 Mutually Exclusive Events

All the events discussed so far in this chapter are *mutually exclusive*. If a die shows two spots face up, then it will not show three or six spots in the same roll. Now, suppose someone decides to alter the odds by constructing a die whose four sides have one, two, three, and four spots while the two remaining sides have five spots. What is the probability of throwing a 5 this time? The player has two out of six chances of throwing a 5.

The probability that an event will occur is the sum of the probabilities of the mutually exclusive events of which it is composed.

What is the probability of throwing an odd number, a 1, a 3, or a 5, with an ordinary six-sided die? The probability is three out of six or one half.

What is the probability of drawing a spade from a deck of 52 cards? A deck has 13 spades, so the odds are 13/52 or one fourth.

Two thousand African-Americans live in my predominantly Caucasian court district of 100,000 people. What is the chance the next prospective juror summoned will be an African-American? Two thousand chances out of 100,000.

If the last 10,000 prospective jurors are Caucasian, what is the probability the next juror summoned will be African-American? Two thousand out of 100,000? Not quite. Local court rules specify that once you are summoned, whether or not you serve on a jury, you cannot be called for two more years. That means the correct answer is 2000 out of 99,000.

To determine the probability P of a specific event, we need to know both N, the number of possible outcomes, and n, the number of mutually exclusive outcomes that will lead to the event of interest.

Probability P equals n/N.

5.2.1 Which Population?

Often we will want and need to apply probabilities to populations that are not as well defined as in the preceding examples. For example, we may want our expert to testify to the probability that lightning will strike at a particular place or that an individual might contract leukemia even in the absence of some specific traceable cause. Our population consists of successive slices in time (a second, a month, a year). When our expert testifies that the probability of lightning striking a certain spot on a particular day is one in 10,000,000, she means that if conditions were to remain the same over a period of 10,000,000 days, on at least one of those days, she would expect to observe lightning striking the ground in the designated area. If questioned as to how she arrived at that figure, she might reply that she had examined the records of 1000 comparable areas over a period of 10,000 days. Ensuing arguments could then focus, and properly so, on whether the various areas were indeed comparable, that is, equally likely, whether days in spring should have been mixed in with days in summer, and so forth.[4]

5.2.2 Putting the Rules in Numeric Form

Statisticians are really mathematicians who have taken a wrong turn. Like accountants, they want to see all their results expressed in numerical, quantitative terms. Rather arbitrarily, they have decreed that if A is an event of interest, and P(A) is the probability an event will occur, then P(A) cannot be smaller than zero or larger than one: $0 \leq P(A) \leq 1$. If A is certain to occur, then $P(A) = 1$. If A will never occur, then $P(A) = 0$.

Suppose an event A is made up of two mutually exclusive events, B and C; that is, A occurs if either B or C occurs. An example would be A: the stolen diamond is in Smith's coat pocket; B: Smith put the diamond there; and C: someone else put the diamond there. Then $P(A) = P(B) + P(C)$. Of course, the probability that B and C occur simultaneously is 0 since B and C are mutually exclusive; therefore, $P(B \text{ and } C) = 0$.

Recall from your lessons in logic that if A does not occur, then its opposite, the event notA, must occur and vice versa: $P(\text{notA}) + P(A) = 1$ or $P(\text{notA}) = 1 - P(A)$.

5.3 Conditional Probabilities

During the 1970s, the company I worked for decided to hire a larger number of African-Americans than it had in the past, placing the majority of them in entry-level jobs. By the 1980s, the same African-Americans who started delivering mail five or six years earlier were still at work in the mailroom. They had been denied the opportunities for advancement afforded other employees. The mere ratio of African-Americans to Caucasians employed by the company did not tell the entire story.

In California in the early 1990s, about 10% of community college faculty members were African-American, a percentage that more or less mirrored the percentage of African-Americans in the population at large. Twenty-five percent of the administrators (deans, assistant deans, college presidents, and vice presidents) were African-American. Was the percentage reasonable? Apart from a few appointed positions, most community college administrators are drawn from the faculty. Investigating reverse discrimination, we are

---

[4] This frequentist view of probability is due to von Mises [1928, 1957].

interested in the conditional probability that an African-American who has joined the faculty will go on to be promoted to an administrative position. It is obvious that the *conditional probability* of this event is much greater than it would be for a Caucasian faculty member.

Another example deals with the selection of jurors. Suppose that, in accordance with the law, the composition of jury panels reflects that of the population at large, but that African-Americans, Caucasians, or Hispanics are mysteriously absent from the juries. We then would be concerned with the conditional probability that an individual summoned for a jury panel will subsequently serve on a jury.

If a child has a certain gene, then the probability he inherited it from his mother rather than his father is one out of two or 50%. If we test the father and learn he does not have the gene, the *conditional probability* that the mother has the gene, given the new information, is 100%.

The Daily Double at the Santa Anita race track is based on a simple requirement. All you have to do is to pick the winners of the seventh *and* the eighth races. Occasionally someone will buy a ticket with no real expectation of success. When he learns his horse has won the seventh race, his palms sweat and his heart pounds until he sees which horse crosses the finish line in the eighth race. Here is why: the original probability of winning the daily double was less than one in 100. Based on the success of the horse in the seventh race, the conditional probability of winning the daily double became one in ten, maybe as favorable as one in three if the bettor is a clever handicapper.

5.3.1 Negative Evidence

Negative evidence can be revealing. Recall Sherlock Holmes and the case of the dog that did *not* bark in the night. The Monte Hall paradox involves three doors marked A, B, and C. A complete set of matched luggage and two round tickets to Tahiti lie behind one of the doors. You choose door C. You have no particular feeling about it, and despite the screams from the audience, one door seems as good as another because the probability formula is $\Pr\{C\} = 1/3$.

Monte asks you if you want to switch. You shake your head. He opens door A, revealing an empty cupboard and asks you again. What should you do?

One could argue that the two remaining doors were equally likely to begin with. They are still equally likely, that is $\Pr\{C\} = 1/2$ and, despite the continuing screams from the studio audience, there is no point in switching. This interpretation ignores the negative evidence.

Before door A was opened, $\Pr\{A \text{ or } B\} = 2/3$. The conditional probability of B given that A did not contain the prize, $\Pr\{B \mid \text{not } A\}$, is thus 2/3 also. Switch. The odds are on your side.

Gambler's Ruin

Must a run of bad luck be followed by good? Not if the events are independent of one another as the following case illustrates:

Heather Devon plunked dollar coins for 12 hours straight into a slot machine at the Frontier Hotel-Casino under the totally unfounded belief that she was bound to hit the jackpot sooner or later. The casino staff promised to lock down the machine for her while she ate breakfast, then reopened it for another player in return for a $20 tip.

The new player won a $97,823 jackpot and Heather sued the casino alleging the jackpot should have been hers. Trial testimony by an employee of International Gaming Technology of Reno, which built the machine, revealed the following:

■ The position in which a slot machine's reels stop is based on a series of random, time-dependent, constantly changing numbers picked by the computer inside the machine.
■ Successive picks are independent of one another.

Past behavior of the machine was no guide to future behavior; to win the jackpot, Devon would have had to pull the lever at the exact same millisecond as the winning player.

5.4 Independence

The preceding discussion of conditional probability serves as a prelude to an equally important concept, that of *independence*. Events A and B are said to be independent if the conditional probability of the one given the other, $P(A|B)$, is the same as its unconditional probability $P(A)$; that is, a knowledge of whether B occurred is of no value for predicting A. If you flip a coin twice in a row, the probability it will come up heads the second time is the same, whether or not it comes up heads on the first throw. The two events — heads on the first trial and heads on the second — are completely independent. On the other hand, the probability of winning the daily double depends totally on choosing the horse winning the seventh race.

To justify the results of a sample or a survey to the court, you must show that your observations are independent of one another. Suppose you did a survey of political preferences by going from house to house and interviewing the occupants while they sat around the dinner table. Their opinions would not be the same as those they would express in the privacy of a polling booth. People care what others think and, when asked for an opinion about an emotionally charged topic in public, may or may not tell the truth. In fact, people are unlikely to tell the truth if they think third parties are listening. If a wife hears her husband say he is a Democrat, she may reply that she is a Democrat in order to avoid an argument. Their responses might differ but more often than not will depend on what one thinks the other wants to hear. Spouses are more likely to have similar political views than, say, next-door neighbors. Neighbors are more likely to have similar views than two individuals selected completely at random because the neighbors probably belong to the same socioeconomic category. If I know what one neighbor believes, I can probably make a pretty good, but not perfect, guess as to another neighbor's feelings on a particular issue. Their answers are dependent, though to a lesser extent than the answers of two spouses.

The two race track events: A — my horse wins the seventh race, and B — I win the daily double, are dependent because $P(B|A) > P(B)$.

5.4.1 The Product Rule

If A and B are any two events, then:

$$P(B) = P(A) * P(B|A) + P(notA) * P(B|not A)$$

As will be seen in the next chapter, this product rule has a poor reputation in the criminal courts. Nonetheless, it has many practical applications. At the race track, let A be *my horse wins the seventh race.* B is *I win the daily double.* $P(B) = P(A) * P(B|A)$; that is, if my horse does not win the seventh race (notA), I will not win the daily double. C is *my horse wins the eighth race.* Then:

$$P(B) = P(A \text{ and } C) = P(A) * P(A \text{ and } C|A) = P(A) * P(C|A) = P(A) * P(C)$$

For the outcomes of the seventh and eighth races, A and C are clearly independent.

Here is a practical application. Bet on the daily double only if it yields a better payoff than betting on the seventh race and reinvesting all your winnings (if any) to bet on the eighth race.

5.4.2 DNA Matching

The simple product rule is often used to estimate probabilities used in DNA matching. For example, suppose 10% of the sperm in the gene pool carry allele 1 (A_1), and 50% carry allele 2 (A_2). Similarly, 10% of the eggs carry A_1, and 50% carry A_2. If individuals mate at random, without regard to the presence or absence of allele A_1 and A_2, we can expect 5% of the fertilized eggs to be (A_1, A_2) and another 5% to be (A_2, A_1). Both configurations produce identical autoradiograms with one band for A_1 and another band for A_2. The expected proportion of heterozygotes $A_1 A_2$ is 5% + 5% = 10%.

Table 5.1 Combining Genotypes

Eggs	Sperm	
	Allele 1 (10%)	Allele 2 (50%)
Allele 1 (10%)	10% × 10% = 1%	10% × 50% = 5%
Allele 2 (50%)	50% × 10% = 5%	50% × 50% = 25%

The sum of the entries in this table is 0.6 * 0.6 = 36% rather than 100% because we are not interested in individuals who may carry some other allele.

More generally, when the frequency of two alleles is p_1 and p_2, the single-locus genotype frequency for the corresponding heterozygotes in a randomly mating population is expected to be 2 p_1 p_2. The single-locus genotype frequency for the corresponding homozygotes is expected to be p_{12} and p_{22} (1% and 25% in our table). These proportions are known as Hardy–Weinberg equilibrium proportions. For more on this topic, including the consequences of nonrandom mating, see Kaye [1997].

5.4.3 Sampling with and without Replacement

We can put these rules of probability to work again on some common problems. In Chapter 2, we noted Justice Blackmun's comment in *Ballew v. Georgia*[5] that if a minority group comprises 10% or less of a population, a jury of 12 persons selected at random from that population will fail to include members of that minority at least 28% of the time.

---

[5] 435 U.S. 223 (1978).

If the population in question is large enough, say, 100,000 or so, we can assume that the probability of selecting a nonminority jury member is a constant 90 out of 100. The probability of selecting two nonminority persons in a row according to the product rule for independent events is 0.9×0.9 or 0.81. Repeating this calculation ten more times, once for each of the remaining ten jury persons, we get a probability of $0.9 \times 0.9 \times \ldots \times 0.9 = 0.28243$, as Justice Blackmun specified.

Of course, if the percentage of minorities in the population rose to 20%, we would expect an even lower percentage of juries would fail to include minorities. How much lower? Based on $0.8 \times 0.8 \times \ldots \times 0.8$, the answer is 0.06879 or about 7%.

Assume a new set of jury commissioners put together a panel of 40 Caucasians, 5 African-Americans, and 5 Hispanics. This seems fair because the proportions of the various races are identical with those in the population at large. That is, it seems fair until the jury for our case includes no non-Caucasians. Events like this happen 7% of the time. Or do they? Successive selections from a panel with only a small number of members will not be as independent as they would be if drawn from a very large population. Every time an individual is taken from a panel and put on a jury, the composition of the panel changes. The actual probability of selecting a jury of 12 Caucasians from our hypothetical panel is 4.6% or:

$$(40/50) * ([40 - 1]/[50 - 1]) * ([40 - 2]/[50 - 2]) * \ldots * ([40 - 11]/[50 - 11])$$

5.5 Bayes' Theorem

Since its communication to the Royal Society in 1763,[6] Bayes' theorem has exerted a near fatal attraction on those exposed to it.[7] I hesitate to award it even passing mention here. Much as a bell placed on a cat magically resolves so many of the problems of the average house mouse, Bayes' straightforward, easily grasped mathematical formula would appear to provide a long-awaited tool for a robotic judge free of human prejudice.

Suppose we have a set of evidence $E = \{E_1, E_2 \ldots E_n\}$ and have determined the conditional probability $Pr\{A|E\}$ that some event A is true. A might be that O.J. Simpson killed his ex-wife, that the captain of the Exxon Valdez behaved recklessly, or some other incident whose truth or falsehood we wish to establish. An additional piece of evidence E_{n+1} now comes to light. Bayes' theorem tells us that

---

[6] *Phil. Tran.* 1763; 53:376-398.
[7] The interested reader is directed to Keynes [1921] and Redmayne [1998].

$$Pr\{A|E_1, ..., E_n, E_{n+1}\} =$$

$$\frac{Pr\{E_{n+1}|A\}Pr\{A|E_1, ..., E_n\}}{Pr\{E_{n+1}|A\}Pr\{A|E_1, ..., E_n\} + Pr\{E_{n+1}|{\sim}A\}Pr\{{\sim}A|E_1, ..., E_n\}}$$

where ~A (read notA) is the event that A did not occur. Recall that Pr{A} + Pr{~A} = 1. Pr{A | E_1, ..., $_n$ is the *prior* probability of A, and Pr{A | E_1, ..., E_n, E_{n+} the *posterior* probability of A once the item of evidence E_{n+1} is in hand. Gather sufficient evidence and you shall gain an automatic verdict.

The problem with the application of Bayes' theorem in practice comes at the beginning of a trial when you have no evidence in hand, and n = 0. What is the prior probability of A then? How the courts have dealt with this problem is covered in Sections 6.1 and 7.2.3.

5.6 Summary

If N is the total number of mutually exclusive possible outcomes and n is the smaller number of mutually exclusive outcomes that will lead to a particular event A, then:

$$0 \le P(A) = n/N \le 1$$

$$P(notA) = 1 - n/N$$

If B and A are mutually exclusive events, then P(A or B) = P(A) + P(B); P(A and B) = 0. If A and B are any two events, then:

$$P(B) = P(A) * P(B|A) + P(notA) * P(B|notA)$$

If A and B are two independent events then P(B|A)=P(B) and P(A and B) = P(A) * P(B); this last is often referred to as the product rule.

5.7 To Learn More

Zeisel and Kaye [1997] contains an extensive and lucid analysis of the courts' treatment of DNA evidence. Probability calculations needed to support forensic evidence are provided by Aitken [1995]. Unfortunately, as we shall see in the next chapter, few criminal courts are willing to accept them.

Chapter 6

Criminal Law

Statistical calculations rival a polygraph in their unreliability and propensity to mislead.[1]

Mathematical odds are not admissible as evidence to identify a defendant in a criminal proceeding so long as the odds are based on estimates whose validity has not been demonstrated. The laws of probability cannot be used to combine distinct items of circumstantial evidence. On the other hand, probabilities can be used to establish specific pieces of evidence — the rarity of a blood sample or a DNA fragment — but only when confimed by extensive experimental or survey evidence. Expert opinion simply is not enough.

6.1 Facts versus Probabilities

The courts have held it reversible error to permit an expert witness to testify as to mathematical probabilities that show the defendant was the person who committed the crime.

In *State v. Sneed*,[2] the state of New Mexico attempted to prove the defendant purchased a handgun on the morning of the murder. Witnesses testified that the defendant used the name Robert Crosset at two places one week before the night of the murder, and that a person by that name

---

[1] *Chumbler v. Commonwealth*, 905 SW.2d 488, 495 fn. 17 (1995, Ky).

[2] 76 NM 349, 414 P.2d 858 (1966).

purchased a handgun from a pawnshop on the morning of the murder. The pawnshop register read "Robert Crosset, Box 210, Las Cruces, 5 ft. 9 in., brown hair and brown eyes." A professor of mathematics, called as a witness by the state, was firmly grounded in the principles enunciated in the preceding chapter. He testified that if you have several events that are independent, estimate the probability of each of those events, and multiply those numbers together, you have a good idea of the probability that all of them will happen.

For one set of numbers, the witness examined telephone books from various western communities, estimated the books contained 1,290,000 names, and found that the *Crosset* name did not appear. The *Robert* name appeared about once in every 30 names. Combining the two results, the witness' estimate of probability was that *Robert Crosset* would appear once in 30 million names.[3] For his second set of numbers he used the store register, which was in evidence. He testified that it contained 35 listings and 12 of the entries were "brown-brown," (the colors of eyes and hair). He stated that 12 of the 35 entries showed heights between 5 ft. 8 in. and 5 ft. 10 in. He further testified that the probability that two people at random would choose the same post office box number from 1000 numbers was one in 1000. After estimating these probabilities, the result of applying the multiplication rule for independent probabilities was one in 240 billion. The significance of this figure, according to the witness, was a chance of 240 billion to one that the defendant was responsible for the series of facts in the pawnshop register, as opposed to another person who entered the pawnshop and accidentally implicated him. "This," said the witness, is the application of "this thing [probability] to criminalistics."

In holding that it was error to admit this testimony in evidence, the appellate court pointed out that the witness explained how his estimates were made and used conservative estimates, but he also testified that, "One might argue about these numbers and what they mean." When asked if he were testifying "that these are incontrovertibly the chances that were present," his answer was "no." The court further observed that the witness did not testify as to why he chose these particular items on which to base his estimates, nor did he state why a positive number was used in arriving at an estimate on the basis of the telephone books when Robert Crosset was not listed in those books.[4] The court concluded that mathematical odds are not admissible as evidence to identify a defendant in a criminal proceeding if the odds are based on estimates whose validity has not been demonstrated. The defendant was granted a new trial.

---

[3] The opinion does not reveal how he arrived at this estimate or whether he stated and was able to show that first and last names are independent.

[4] This procedure yields an estimate that most favors the defendant.

People v. Collins[5] concerned the prosecution of an African-American and his Caucasian wife for robbery. The victim testified that her purse was snatched by a girl with a blond ponytail. A second witness testified that he saw a blond girl, ponytail flying, enter a yellow convertible driven by an African-American male with a beard and mustache. Neither witness could identify the suspects directly. In an attempt to prove the defendants were in fact the persons who had committed the crime, the prosecutor called a college instructor of mathematics to establish that, assuming the robbery was committed by a Caucasian female with a blond ponytail who left the scene in a yellow Lincoln accompanied by a African-American with a beard and mustache, there was an overwhelming probability the crime was committed by any couple who had such distinctive characteristics.

In substance, the witness testified to the product rule,[6] which states that the probability of the joint occurrence of a number of mutually independent events is equal to the product of the individual probabilities that each of the events will occur. Without presenting any statistical evidence in support of the probabilities for the factors selected, the prosecutor proceeded to have the witness assume probability factors for the various characteristics that he deemed to be shared by the guilty couple and all other couples answering to their distinctive characteristics. Applying the product rule to his own factors, the prosecutor arrived at a probability of one chance in 12 million that a couple possessed the distinctive characteristics of the defendants. Under this theory, it was to be inferred that there could be only one chance in 12 million the defendants were innocent and that another equally distinctive couple had actually committed the robbery.

Expanding on what he had suggested as a hypothesis, the prosecutor offered what the court described as the completely unfounded and improper testimonial assertion that, in his opinion, the factors he assigned were conservative estimates and that, in reality, the chances that anyone other than the defendants was at the scene and had every similarity approached one in a billion. On appeal, the defendants contended that the introduction of evidence pertaining to the mathematical theory of probability and the use of it by the prosecution during the trial constituted prejudicial error.

In reversing the defendants' conviction and ordering a new trial, the court said that the prosecution's introduction and use of mathematical probability statistics injected two fundamental errors into the case:

---

[5] 68 Cal. 2d 319, 66 Cal. Rptr. 497 (1968).

[6] Discussed in Section 5.4.1.

1. The testimony lacked an adequate foundation both in evidence and in statistical theory.
2. The testimony and the manner in which the prosecution used it distracted the jury from its proper and requisite function of weighing the evidence on the issue of guilt; encouraged the jurors to rely upon an engaging but logically irrelevant expert demonstration; foreclosed the possibility of an effective defense by an attorney apparently unschooled in mathematical refinements; and placed the jurors and defense counsel at a disadvantage in sifting relevant fact from inapplicable theory.

From a statistician's point of view, the prosecutor's errors are twofold. First, estimates were used instead of facts. The prosecutor suggested the odds were one in four that a girl in San Pedro would be blond — but he provided absolutely no facts in support of this simple allegation — not even a survey of local hairdressers. Similarly, he "suggested" that only one in ten of the cars in the area was yellow, again without any supporting evidence. To quote Justice Sullivan:

> We seriously doubt that such evidence could ever be compiled since no statistician could possibly determine after the fact which cars, or which individuals "might" have been present at the scene of the robbery; certainly there is no reason to suppose that the human and automotive populations of San Pedro California, include *all* potential culprits — or, conversely, that all members of these populations are proper candidates for inclusion.[7]

Second, the prosecutor's probability calculations were in error. As we saw in Section 5.5.1, the product rule applies only if the events are independent. African-Americans with beards and African-Americans with moustaches represent overlapping categories, although they were treated as separate and independent by the prosecution's expert witness. Beards, yellow convertibles, and a girl friend of a different race all seem associated with a ceratin type of flamboyant personality rather than being independent traits. The appeals court, commenting on the failure of the witnesses to make a positive identification, wondered whether, "the guilty couple might have included a light-skinned Negress with bleached hair rather than a Caucasian blond; or the driver of the car might have been wearing a false beard as a disguise ...."[8]

---

[7] Id. at n. 12. Italics are the author's.

[8] We find a similar problem with the independence assumption in *State v. Sneed*: are the names *Robert* and *Crosset* really independent?

6.1.1 Exception to the Rule

While the reasoning in *People v. Collins* prevails almost universally today, a quite different result was reached in Georgia in a murder case that attracted nationwide attention. Wayne Williams was convicted of the murders of two young African-American males in Atlanta.[9] The evidence against him included fibers found on the bodies that were similar to the fibers in a carpet in Williams' bedroom. A prosecution expert testified that manufacture of this type of carpet fiber had been discontinued. He "guesstimated" only enough fiber to carpet 820 rooms was sold in a ten-state area and suggested without proof that (1) if sales were equal in each state, (2) if all the carpet allotted to Georgia was sold in Atlanta, and (3) if only one room per house was carpeted, the odds were approximately 1 in 8000 that a home selected at random in the Atlanta area would contain fibers similar to those found on the bodies of the murdered boys. The Georgia court of appeals upheld the use of the expert's testimony.[10]

6.1.2 Bayes' Theorem

Bayes' theorem has seen little use in criminal trials as it ultimately relies on estimates and thus is subject to the same objections enumerated in *People v. Collins.* Tribe [1971] states several objections including the argument that a jury might actually use the evidence twice, once in its initial assessment of guilt — that is, to determine a prior probability — and a second time when it applies Bayes' theorem. A further objection to its application is that if a man is innocent until proven guilty, the prior probability of his guilt must be zero. By Bayes' theorem, the posterior probability of his guilt would be zero also, rendering a trial unnecessary. The courts of several states have remained unmoved by this argument.[11]

In *State v. Spann,*[12] showing the defendant fathered the victim's child was key to establishing a charge of sexual assault. The state's expert testified that only 1% of the presumed relevant population of possible fathers had the same type of blood and tissue that the father had and, further, that the defendant was included within that 1%. In other words, 99% of the male population at large was excluded. Next, the expert used

---

[9] You may recall there had been ten similar murders.

[10] *Williams v. State*, 252 Ga. 749, 312 S.E. 2d 40 (1983), citing *Stewart v. State*, 256 Ga. 70, 75, 268 S.E. 906 (1980), and *Wisdom v. State*, 234 Ga. 650, 655, 217 S.E. 2d 244 (1975).

[11] See, for example, *Davis v. State*, 476 N.E. 2d 127 (Ind. App. 1985) and *Griffith v. State of Texas*, 976 S.W. 2d 241 (1998).

[12] 130 N.J. 484 (1993).

Bayes' theorem to show that the defendant had a posterior probability of fathering the victim's child of 96.5%.

> The expert testifying that the probability of defendant's paternity was 96.5% knew absolutely nothing about the facts of the case other than those revealed by blood and tissues tests of defendant, the victim, and the child ....[13]

> In calculating a final probability of paternity percentage, the expert relied in part on this 99% probability of exclusion. She also relied on an assumption of a 50% prior probability that defendant was the father. This assumption, [was] not based on her knowledge of any evidence whatsoever in this case ... [she stated] everything is equal ... he may or may not be the father of the child.[14]

> Was the expert's opinion valid even if the jury disagreed with the assumption of .5 [50%]? If the jury concluded that the prior probability is .4 or .6, for example, the testimony gave them no idea of the consequences, no knowledge of what the impact (of such a change in the prior probability) would be on the formula that led to the ultimate opinion of the probability of paternity.[15]

> The expert's testimony should be required to include an explanation to the jury of what the probability of paternity would be for a varying range of such prior probabilities, running for example, from .1 to .9.[16]

Courts in California,[17] Illinois, Massachusetts,[18] Utah,[19] and Virginia[20] also challenged the use of the fifty-fifty assumption. The notion of providing a range of probabilities has been echoed in a number of civil trials (see Section 7.2.4). In *State v. Jackson*,[21] the expert included a range of prior probabilities in her testimony, but the court ruled the trial judge had erred in allowing the expert to testify as to the conclusions of Bayes'

---

[13] Id. at 489.
[14] Id. at 492.
[15] Id. at 498.
[16] Id. at 499.
[17] *State v. Jackson*, 320 NC 452, 358 S.E. 2d 679 (1987).
[18] *Commonwealth v. Beausoleil*, 397 Mass. 206 (1986).
[19] *Kofford v. Flora*, 744 P. 2d 1343, 1351-1352 (1987).
[20] *Bridgeman v. Commonwealth*, 3 Va. App. 523 (1986).
[21] 320 N.C. 452 (1987).

theorem in stating a conclusion that the defendant was "probably" the father of the victim's child.

6.2 Observations versus Guesstimates

We need not forget all we learned about probability in Chapter 5 and forgo the use of forensic evidence including blood samples and dental records. We simply need to make the same distinction the courts make and distinguish speculation and "guesstimates" from well-grounded observation.

In a murder trial, the lower court permitted the prosecution to establish from sperm and blood samples associated with the defendant and crime that the defendant and whoever committed the crime shared characteristics found only in a small percentage of the general population. "These computations were neither misleading nor confusing and the prosecution did not attempt to reduce the ultimate question of innocence or guilt to one of mathematical probabilities."[22]

In *People v. Slone*,[23] a major portion of the prosecution's case against the defendant was the presentation of testimony from three expert witnesses, dentists, who made a positive comparison between the defendant's dentition and a bite mark on the victim's thigh. As the court said in upholding its admissibility:

> There is no merit to defendant's corollary contention that by employing screening of thousands of cases at the U.C.L.A. Dental Clinic, the experts were attempting to impose mathematical probability statistics or odds on the fact-finding process …. The experts in the instant case were simply attempting to negate the potential disapproval of their scientific method in the area of specificity — the problem posed by the defense counsel, who inquired whether the experts could testify that no other human being on the planet could have bit the victim on the thigh. The expert witnesses were careful to say that they could not. There is a probability factor in even the most carefully structured scientific inquiry; seldom is it possible to exclude all possible chance for error in human endeavor. But there is no requirement in our law that the admissibility of scientific test evidence must be predicated on a 100% degree of accuracy.[24]

---

[22] *U.S. v. Gwaltney*, 790 F.2d 1378 (9th Cir. 1986), 20 Fed. Rules Evid Serv. 1293, cert. denied, 479 U.S. 1104. See also *Scott v. Perini*, 622 F. 2d 428, 430 (6th Cir. 1981), cert. denied, 456 U.S. 909 (1982).

[23] 76 Cal. App. 3d 611.

[24] Id. at 625.

In another murder trial, a dentist testified to an eight in one million probability that teeth marks found on a decedent's body were not made by the defendant. His testimony was found admissible because it derived not from personal mathematical calculation but from articles in medical journals and books providing numerical values for finding two sets of teeth of the same type.[25]

A mathematics professor testified in *People v. Risley*.[26] As in *Garrison*, the court said the statement of the witness was not based on actual observed data, but was simply speculative; but the court distinguished the evidence the professor offered from other statistical evidence such as life expectancy tables based on actual observation and data.

A Massachusetts court ruled that testimony by a hair chemist that only one person in 4500 would have hair with the same characteristics as a hair fragment recovered from a victim's shirt is properly admitted where such testimony has a sound foundation and is based on fact and not conjecture.[27]

DNA evidence played a major role in convicting Lynda Axell of first degree murder.[28] She appealed the verdict on several grounds, among them, whether the basis for the calculation of statistical probability employed by the testing laboratory, Cellmark, satisfied the foundation requirements of *People v. Collins*.

> In *Collins*, unlike in the instant case, the technique to measure probabilities suffered from two basic and pervasive defects — an inadequate evidentiary foundation and an inadequate proof of statistical independence which is essential to the proper application of the "product" or "multiplication" rule.[29] Respondent concedes the need for some form of statistical evidence since a match between two DNA samples means little without data on the probability of the match having occurred between two random individuals.

> Here defense witnesses testified that calculation of the probability of a random match in a population depends upon four major assumptions: (1) the correct population has been identified; (2) the population sample is large enough that the observed frequencies accurately represent the true population frequencies; (3) the sample is truly random; (4) the population

---

[25] *State v. Garrison*, 120 Ariz. 255, 585 P.2d 563, 566, 568 (1978).

[26] 214 N.Y. 75, 108 N.E.200 (1915).

[27] *Commonwealth v. Hyatt*, 557 N.E.2d 1172 (1990), review granted, 408 Mass. 1104, 562 N.E.2d 90; superseded on other grounds, 409 Mass. 689, 568 N.E.2d 1148.

[28] *People v. Axell*, 235 Cal. App. 3d 836.

[29] 68 Cal.2d 319, 327, 328.

is homogeneously mixed, in the technical sense that each locus is in Hardy–Weinberg equilibrium and the loci are together in linkage equilibrium.

Appellant asserts that Cellmark failed to identify the correct population with which to identify her. The population to which her alleles were compared was "Hispanic" which is described as one of Spanish surnames or of Spanish descent and could include blacks, whites, Filipinos, Mexicans, Spanish, and any number of subpopulations of South, Central or North America.

Doctor Taylor [a defense expert witness] said the data base fails to include individuals of native American or South American ancestry and that single locus genotypes for appellant are underrepresented in the data base. Doctor Geisser [a second defense expert witness] thought the population sample for which the allele frequency have been derived is not large enough to accurately report the true population frequency. However, appellant and her family characterized her background as "Hispanic," which is a social or geographic term rather than genetic. Since an Hispanic data base may include those of American Indian ancestry, use of an Hispanic data base from Southern California was proper.

Doctor Forman [a prosecution expert witness] testified that she spoke with the person in charge of sending the blood samples from Los Angeles and assured herself that he was familiar with statistics and understood the importance of random sampling. Doctor Kidd [a second prosecution expert witness] also opined that the data base was representative of Hispanics in Southern California. Doctor Forman acknowledged that there appeared to be an excess of single band patterns in the Hispanic data base. However, she testified that she agreed with many experts that calculations can be derived from the data presently available, regardless of their conformance to Hardy–Weinberg equilibrium. She also stated that, based upon discussions with Doctor Kidd, Doctor Conneally, and Doctor Eric Lander about the possible deviation of genotype predictions based on the likelihood of linkage disequilibrium, Cellmark can rely on information in the Hispanic data base to calculate gene frequencies without knowing whether or not there is linkage disequilibrium. She also disputed Doctors Mueller and Taylor because they did not completely analyze their data set.

[B]oth California and the majority of other jurisdictions have traditionally admitted statistical blood-group evidence of this kind in criminal cases, even where it simply includes the accused within the class of possible donors.[30]

Some jurisdictions, as respondent points out, have considered that questions concerning contamination of a sample, chain of custody, reliability of particular results, as well as the size or ratio of the population frequency and statistical probabilities relate to the weight of the evidence and not its admissibility.[31] In *People v. Collins, supra*, the question of statistical probabilities was not a *Kelly/Frye* issue, but one of lack of foundation.

Two issues emerge: (1) are the methods used in calculating the statistical probabilities ones scientifically accepted as generally reliable in the particular field and (2) is there evidentiary support for the particular application? We find that since a match between two DNA samples means little without data on probability, the calculation of statistical probability is an integral part of the process and the underlying method of arriving at that calculation must pass muster under *Kelly/Frye*. However, the size or ratio of the population frequency is a matter of weight rather than admissibility. As in *People v. Castro*,[32] and *People v. Collins*,[33] where the results are so unreliable or completely lack evidentiary foundation, they are inadmissible as a matter of law.

In *Com. v. Curnin*,[34] the court held that there is no demonstrated general acceptance or inherent rationality of the process by which Cellmark arrived at its conclusion that one Caucasian in 59 million would have the DNA components disclosed by the test in that case. Doctor Mueller testified in *Curnin* that Cellmark's data base was not adequate and questioned whether significant substructuring, i.e., non-random mating exists within racial groups which would affect probability determinations

---

[30] *People v. Brown, supra*, 40 Cal.3d 512, 536, fn. 6; accord, *People v. Coleman* (1988) 46 Cal.3d 749, 778-779, fn. 23 [251 Cal. Rptr. 83, 759 P.2d 1260]; *People v. Yorba* (1989) 209 Cal. App. 3d 1017, 1026 [257 Cal. Rptr. 641]; *People v. Morris, supra*, 199 Cal. App. 3d 377, 391.

[31] See, for example, *State v. Pennington, supra*, 393 S.E.2d 847; *People v. Castro, supra*, 545 N.Y.S.2d 985, 999; *People v. Yorba, supra*, 209 Cal. App. 3d 1017, 1026-1027.

[32] *Supra*, 545 N.Y.S.2d 985, 999.

[33] *Supra*, 68 Cal.2d 319, 327.

[34] *Supra*, 565 N.E.2d 440, 442.

using Cellmark's data base. In *Curnin*, however, the prosecution presented no expert to support Cellmark's conclusions. Moreover, the prosecution's expert on Cellmark acknowledged that she was not qualified to give an opinion on the subject.

In *U.S. v. Jakobetz*,[35] the defendant's experts also claimed a lack of factual basis for asserting that substructures or subgroups for the alleles do not exist within the Caucasian race and that the use of the product rule to exhibit genotype frequency is "wholly inappropriate." The court found the defense testimony enlightening but that it did not substantially undermine the FBI genotype frequency procedures as a whole. The court felt that to the extent that substructure might exist, the FBI had sufficiently proved that it compensated for this possibility using conservative binning procedures. The court also noted that recently, "There has been general agreement that Hardy–Weinberg is a poor test for substructuring, at least with the sample sizes involved here. [225 FBI agents.]"

In Jakobetz, as here, Doctor Kidd testified that from looking at data from many subgroups, i.e., Irish, Swedes, Amish, all have "very small differences" in allele frequencies. The court concluded that it is highly unlikely the FBI's frequency estimate of a specific genotype across four or five loci would be lower or prejudicial to the defendant than the actual frequency of that genotype if in fact substructures existed and a less conservative bin system used.

Doctor Kidd testified here that the data base was adequate and acceptable within the scientific community. Moreover, appellant's loci here each displayed a double-band or heterozygous pattern. He opined that the sample is representative of Hispanics in Southern California, the relevant population group. Other courts have recognized that conservative or reduced calculations such as used by Cellmark may correct any Hardy–Weinberg deviation problems.[36]

Doctor Conneally testified that it is standard procedure to use blood banks and that there are only three inbred populations

---

[35] *Supra*, 747 F. Supp. 250, 259-261.
[36] See *Caldwell v. State, supra*, 393 S.E.2d 436, 443 and *People v. Castro, supra*, 545 N.Y.S.2d at 993.

in the U.S.: the Mennonites, Amish, and Hussites. Doctor Mueller admitted that there was no evidence that Los Angeles Hispanics were inbred. Cellmark used a separate data base for each of the four probes and the prosecution experts testified that population samples of 272 and 297 were adequate.[37]

Concerning linkage equilibrium, the relative independence of two genes in the population, Doctor Kidd testified that while two of the four single-locus probes used by Cellmark in this case showed slight linkage, the loci were approaching independence. Doctor Conneally explained that although two of the four loci identified by Cellmark's probes did occupy the same chromosome, the alleles were so far apart as to be transmitted independently. "Experimental data indicates that the probes used by Cellmark are independent of one another."[38] Where the evidentiary foundation is adequate and statistical independence of the characteristics at issue adequately proved, objection to statistical conclusions goes to weight rather than admissibility.[39]

Thus, the prosecution showed that the method used by Cellmark in this case to arrive at its data base and statistical probabilities was generally accepted in the scientific community. Any question or criticism of the size of the data base or the ratio pertains to weight of the evidence and not to its admissibility.

6.2.1   Inconsistent Application

The law in any specific case may be merely what the courts say it is. In a recent ruling in Arizona, evidence from DNA testing in the form of a probability that the match shown by testing was simply a random occurrence was held inadmissible, as "such probability calculations were not generally accepted in the scientific community."[40]

The Arizona court of appeals reiterated this dictum in *State v. Boles* in which the defendant Boles was convicted of multiple counts of burglary, kidnapping, sexual assault, sexual abuse, sexual conduct with a minor,

---

[37] See also *Andrews v. State, supra,* 533 So.2d at 850 [a sample of 500]; *Spencer v. Com., supra,* 384 S.E.2d 775 [a sample of 275].

[38] *State v. Pennington, supra,* 393 S.E.2d at 851.

[39] See *People v. Yorba, supra,* 209 Cal. App. 3d at 1026-1027.

[40] *State v. Clark,* 887 P.2d 572, 164 Ariz. Adv. Rep. 68 (Ariz. App. 1994). This ruling suggests the Arizona court sought to apply *Frye* rather than *Daubert.* See Section 8.1 and Zeisel and Kaye [1997].

and child molestation, involving four victims in neighboring apartment complexes. The evidence against Boles included pubic hairs, sneaker prints, and DNA samples linking him to two victims. The appeals court emphasized that, "The state's experts offered opinions to the effect that it was highly unlikely that someone other than defendant was the source of both samples." The state's two experts and the defendant's expert all testified that they had never seen or heard of two unrelated individuals whose DNA profiles matched over five probes. Indeed, the state's principal expert went so far as to say that to find such a match would require a sample size equal to or greater than the world population. The Arizona court of appeals reversed the conviction, holding that testimony tantamount to uniqueness is not only inadmissible, but also so fundamental an error as to require reversal even without an objection to the testimony.[41]

In a murder prosecution in Illinois, the prosecution attempted to establish the identity of a murderer based on blood types. The admission of expert testimony that "the chances of selecting any two people at random from the population and having them accidentally have identical blood types in each one of these factors is less than one in 500," or 0.2%, was found irrelevant, beyond the scope of expert opinion, highly prejudicial to defendant, and likely to have affected the jury's deliberation.[42]

In a recent murder trial in Kentucky, the prosecution claimed one of the defendants shot the victim from the cover of a shed and introduced evidence regarding a cigarette butt found on the ground in the shed after the killing. The prosecution argued the butt was "a virtual signature" of the defendant, an argument based on analysis of saliva found on the butt and a probability derived from the percentage of the population who are A secretors, the percentage of the population that smoked, and the market share of the brand of cigarette smoked by the defendant. This argument, compelling as it may be on TV, was ruled inadmissible; the probability calculations were completely unfounded and in error.[43]

> The statistical calculations rival a polygraph in their unreliability and propensity to mislead and may have convinced jurors of modest analytical ability that no one but Michael could have committed the crime.[44]

In Maryland, testimony by a detective that 75 to 80% of the approximately 1000 armed robbery cases he investigated while on the police

---

[41] *State v. Boles*, 905 P.2d 572 (Ariz. App. 1995).

[42] *People v. Harbold*, 124 Ill. App. 3d 363, 79 Ill. Dec. 830, 464 NE.2d 734, 748 (1st Dist. 1984).

[43] *Chumbler v. Commonwealth*, 905 SW.2d 488 (Ky. 1995).

[44] Id. at 495, fn. 17.

force resulted in convictions was ruled unreliable and irrelevant; it interfered with the jury's basic function of weighing conflicting evidence.[45]

6.2.2 Middle Ground

The courts may also stake out a middle ground. In a Maryland murder trial, the testimony of an expert who compared blood types and enzyme characteristics of blood found on physical evidence to blood samples from the defendant and the victims and concluded that "between one and two out of 10,000 people" would have the same blood characteristics as the defendant for all of the characteristics tested was admissible as the defense counsel's thorough cross-examination and closing argument on the accuracy of statistical evidence prevented any danger of the jury being misled or confused.[46]

In a prosecution for vehicular homicide, a Massachusetts trial court did not abuse its discretion in permitting the state's expert to testify that the probability that the DNA match was a product of chance was 1 in 18,700 because the court weighed the testimony's relevance and possible prejudice; the testimony was appropriate to assist the jury's understanding; and the trial court possessed additional discretion in making this type of determination.[47]

In a burglary prosecution, a criminalist compared blood samples taken from the scene of crime with those on a piece of white cloth discarded by the defendant. The criminalist testified that only one person out of 100 in a Chicano population would have the combination of chemical elements in his or her blood that were found on the samples. The expert admitted she did not take a blood sample from the defendant or know his blood type. She concluded the defendant was Chicano only because his last name was Gonzalez. The Texas court ruled that, while the speculative nature of her probability calculations might affect the weight to be given them, it did not render them inadmissible.[48]

In a Wisconsin prosecution for sexual assault of a minor, the trial court's admission of testimony by the prosecution's expert witness that only 1% of sexual assault claims by children were fabricated was ruled not in error because the defendant had opened the door for such testimony by previously exploring, during testimony of the defendant's expert witness, the issue of the frequency of children's fabrications of such allegations.[49]

---

[45] *Dorsey v. State*, 350 A.2d 665 (Md. 1976). Other judges might have reached quite a different conclusion.
[46] *Massachusetts v. Paradise*, 405 Mass. 141,156, 539 NE.2d 1006 (1989).
[47] *State v. Schweitzer*, 533 NW.2d 156 (1995, S.Dak.).
[48] *Gonzalez v. State*, 643 SW.2d 751 (Tex. App. 4th Dist. 1982).
[49] *State v. Hernandez*, 192 Wis.2d 251, 531 NW.2d 348 (App. 1995).

6.3   Probable Cause

The courts appear to be split on the issue of whether statistically based psychological profiles can be used as a basis either for arrest or for brief investigative detentions. See, for example, *U.S. v. Lopez*,[50] *U.S. v. Mendenhall*,[51] and *Florida v. Royer*.[52]

6.4   Sentencing

The chief objection the criminal courts have to probabilistic arguments is that they may unduly influence a jury. During the sentencing phase, when only a judge will hear the arguments, the courts can afford to be more liberal.

A fire broke out on a sailboat suspected of carrying contraband as the Coast Guard intercepted it.[53] After the fire was extinguished, approximately 22 plastic-wrapped packages were observed floating in the ocean nearby. Only one of the 22 seemingly identical packages was retrieved, but the First Circuit appeals court agreed with the sentencing judge that it was reasonable to assume that the remaining 21 packages each contained the same amount of marijuana as the single package that was retrieved.

In the *Shonubi* case, in Chapter 4 and discussed below, several probabilistic arguments were offered into evidence, including one proposed by the judge. The Second District court of appeals eventually reversed the case, but on quite different grounds.

6.4.1   U.S. v. Shonubi

Federal courts operate under strict sentencing guidelines. For example, a courier found with less than 5 grams of heroin in his possession might receive a sentence of 10 to 16 months. One found with more than 20 but less than 40 grams would receive a sentence of 27 to 33 months.

Charles Shonubi had a total of 427.4 grams of heroin sealed inside condoms in his digestive tract when he was arrested at Kennedy Airport in 1991. He made seven other trips to Nigeria that year, and the district court determined that all had been for smuggling purposes. His sentence was determined by multiplying 427.4 by 8 for a total of 3419.2 grams that, under the federal guidelines, corresponded to 151 months.[54]

---

[50] 328 F. Supp. 1077 (E.D. N.Y. 1971).

[51] 446 U.S. 544 (1980).

[52] 460 U.S. 491 (1983).

[53] *U.S. v. Hilton*, 894 F.2d 485 (1st Cir. 1990).

[54] *U.S. v. Shonubi*, 802 F. Supp. 859 (E.D. N.Y. 1992).

On the first appeal, the Second Circuit ruled that prior case law "uniformly requires specific evidence — e.g., drug records, admissions or live testimony — to calculate drug quantities for sentencing purposes."[55] Concluding that such evidence was not contained in the record and that multiplication of the quantity seized on the night of the arrest by the total number of trips was an inadequate substitute for the required "specific evidence," the court vacated the sentence and remanded the defendant for resentencing.

> On remand, Judge Weinstein conducted an elaborate hearing. He took testimony from a Government expert on statistics, a defense expert on statistics, and a panel of two statistics experts, appointed by the Court pursuant to Fed. R. Evid. 706. He also received reports of heroin quantities seized from 117 Nigerian heroin swallowers arrested at JFK Airport during the same time period that spanned Shonubi's eight trips. In addition, he surveyed the federal judges of the Eastern District to obtain their opinions concerning heroin swallowers. Based on the record made at an extensive hearing, Judge Weinstein then wrote an elaborate opinion of 177 typescript pages to support his finding that Shonubi had carried between 1,000 and 3,000 grams of heroin during the eight trips.[56,57]

> One of the most significant changes effected by the Sentencing Guidelines is the prescription of precisely calibrated punishment for conduct of which the defendant has *not* been convicted.

> Endeavoring to strike a balance between punishing only for the offense of conviction and punishing for all wrongful conduct that could be established at a sentencing hearing, the Guidelines opted for incremental punishment for conduct deemed to be 'relevant' to the offense of conviction.[58] As to such "relevant conduct," the Guidelines then took the extraordinary and totally unprecedented step of punishing the relevant conduct at precisely the same degree of severity as if the defendant had been charged with and convicted of the activity constituting the "relevant conduct."

---

[55] 998 F.2d 84, 89 (2nd Cir. 1993).

[56] *U.S. v. Shonubi*, 895 F. Supp. 460 (E.D. N.Y. 1995).

[57] This and subsequent extracts from the appeal court's opinion are taken from *U.S. v. Shonubi*, 103 F.3d 1085 (2nd Cir. 1997) unless otherwise noted.

[58] U.S.S.G. §1B1.3.

The "relevant conduct" for a defendant convicted of a drug offense includes all of the additional drugs, beyond the quantity in the offense of conviction, that were unlawfully distributed or possessed with intent to distribute either by the defendant personally or by the reasonably foreseeable acts of others in furtherance of a jointly undertaken criminal activity.[59]

A guideline system that prescribes punishment for unconvicted conduct at the same level of severity as convicted conduct obviously obliges courts to proceed carefully in determining the standards for establishing whether the relevant conduct has been proven. We have recognized the need for such care with regard to the basic issue of the degree of the burden of proof. Thus, though the Sentencing Commission has favored the preponderance-of-the-evidence standard for resolving all disputed fact issues at sentencing,[60] we have ruled that a more rigorous standard should be used in determining disputed aspects of relevant conduct where such conduct, if proven, will significantly enhance a sentence.[61]

A similar concern guided our decision on the prior appeal in this case. Aware of the consequences of a relevant conduct finding as to drug quantities, we invoked the rule from prior case law of our Circuit that, we observed, 'uniformly requires *specific evidence* — e.g., drug records, admissions or live testimony — to calculate drug quantities for sentencing purposes.[62]

The "specific evidence" we require to prove a relevant-conduct quantity of drugs for purposes of enhancing a sentence must be evidence that points specifically to a drug quantity for which the defendant is responsible. By mentioning "drug records" and "admissions" as examples of specific evidence we thought it reasonably clear that we were referring to the defendant — *his* admissions and records of *his* drug transactions. And by "live testimony" we were referring to testimony about *his* drug transactions. Judge Weinstein apparently misunderstood our prior opinion to equate "specific" evidence with "direct" evidence, a consequence that, as he pointed out,[63] would preclude all use

---

[59] Id. §1B1.3(a)(1), (2).
[60] U.S.S.G. §6A1.3., p.s., comment.
[61] See *U.S. v. Gigante*, 94 F.3d 53, 56-57 (2nd Cir. 1996) (denying petition for rehearing).
[62] 998 F.2d at 89 (emphasis added).
[63] 895 F. Supp. at 478.

of circumstantial evidence. However, our identification of drug records as one example of "specific evidence" should have dispelled that misunderstanding since such records are a form of circumstantial evidence. If a defendant's drug records reflect drug transactions of a specific quantity, that is circumstantial evidence permitting the inference that the defendant has trafficked in that quantity of drugs.

If some "specific evidence" of quantity is presented, we do not rule out the possibility that evidence of the sort considered by the District Court might be usefully assessed in determining whether the alleged quantity had been established by a preponderance of the evidence. However, if "specific evidence" of the quantity handled by the defendant (or others for whose acts he is responsible) is available, it is not likely that the time and effort required to conduct probability analyses of quantities carried by other drug couriers would be worthwhile.

Though disapproving of our requirement that the relevant-conduct quantity of drugs be based on "specific evidence," the District Court endeavored to apply this requirement. Judge Weinstein acknowledged that we had required "specific evidence" such as drug records, admissions, or live testimony, and identified evidence that he believed met our standard. For "records" he cited "a combination of drug records (including DEA [Drug Enforcement Agency] and Customs Service records) and the records of Shonubi's trial, sentencing hearing, and presentence report." For "admissions" he cited Shonubi's "admissions at the time of his arrest." For "live testimony" he cited "the statistical analysis introduced on remand as well as testimony on the economics of heroin swallowing."[64]

These items of evidence are not "specific evidence" of drug quantities carried by Shonubi on his prior seven trips. We required specific evidence of what Shonubi had done. The DEA records informed Judge Weinstein of what 117 other balloon swallowers from Nigeria had done during the same time period as Shonubi's eight trips. Those records of other defendants' crimes arguably provided some basis for an estimate of the quantities that were carried by Shonubi on his seven prior trips, but they are not "specific evidence" of the quantities he carried.

---

[64] We consider the specific arguments in the next section.

The defendant's distinguished expert on statistics, Michael O. Finkelstein, Esq., correctly informed the District Court that "statistics relating to others would not usually be characterized as specific evidence relating to Shonubi." The experts on the Court's Rule 706 panel rendered the same advice. Though the records of Shonubi's trial, sentencing hearing, and presentence report relate specifically to Shonubi, they do not provide "specific evidence" of the quantities carried on his prior seven trips, any more than they did when these records were before us on the prior appeal. Shonubi's admissions likewise are "specific" as to him, but contain no "specific evidence" of the quantities carried on his prior trips.

Since the Government has now had two opportunities to present the required "specific evidence" to the sentencing court, no further opportunity is warranted, and the case must be remanded for imposition of a sentence based on the quantity of drugs Shonubi carried on the night of his arrest, adjusted only by the previously adjudicated enhancement for obstruction of justice.[65]

6.4.2 Statistical Arguments

The District Court considered several statistical analyses. The Government's expert, Dr. David Boyum, made two analyses, each based on the DEA's report of 117 balloon swallowers from Nigeria who were arrested at JFK Airport during the time period of Shonubi's eight trips. The first analysis calculated how many of the 117 balloon swallowers in the DEA report carried quantities within 13 100-gram ranges from 0 to 1,300 grams.[66] From this classification, Dr. Boyum calculated that the mean net weight was 432.1 grams and the median net weight was 414.5 grams, figures he deemed reasonable to estimate for Shonubi's previous seven trips. Second, Dr. Boyum entered into a computer the weights carried by these 117 smugglers and asked the computer to calculate the sum carried on seven trips, selected at random from the 117. He then asked the computer to repeat this random selection and calculation 100,000 times.[67] From this process he

---

[65] 103 F.3d 1085 at 1092.

[66] 895 F. Supp. at 500 (Table 1).

[67] He bootstrapped, in other words. See Section 4.2.2.

determined there was a 75 percent probability that Shonubi carried more than 2712.6 grams on his prior seven trips.[68]

The District Court's Rule 706 panel of experts submitted two analyses of their own. As a first step, one of the Rule 706 experts, Prof. David Schum, distributed a pound of powdered sugar into 103 balloons, and also "reflect[ed] on the task of swallowing them." There is no indication that he carried his investigation to the point of swallowing the balloons. Dr. Schum concluded that the activity of carrying heroin in swallowed balloons involves a learning curve. Next, he constructed two charts, each reflecting quantities Shonubi might have carried on his eight trips. For the first trip he assumed the amount was the smallest amount carried by any of the 117 smugglers from the DEA report. For the last trip he used the quantity Shonubi carried when arrested. The first chart estimated the intervening trips by using an arithmetic progression (increasing the quantities in equal intervals). The second chart used the same quantities for the first and last trips, but estimated slightly smaller quantities than the first chart for the intervening trips, to reflect a slower learning curve. The aggregate quantity for the eight trips was 1,930 grams from the first chart and 1,479 grams from the second chart.

Judge Weinstein also constructed his own "non-Bayesian and non-statistical model." First, he estimated the probabilities that Shonubi carried heroin on his eight trips. He used 99 percent for the eighth trip, 95 percent for the seventh trip, and decreased the probability by five percentage points for each of the prior trips. Then, he estimated a range of the quantities carried on each trip. These ranges included the quantity recovered from the last trip, but estimated a bottom of the range that diminished with each earlier trip. He then multiplied the estimated probability by the estimated quantity, using the bottom of the estimated ranges in order to "favor the defendant." Finally, he aggregated three different total quantities by including only those quantities from trips where the estimated probability of carrying heroin exceeded a level that Judge Weinstein associated with different burdens of proof — beyond a reasonable doubt, 95 percent+ probability; clear and convincing, 70 percent+ probability; and preponderance, 50 per-cent+ probability. This yielded total quantities of 752 grams for the two trips with at least a 95 percent probability of carrying

---

[68] 103 F.3d 1085, n. 31t, 1091 (2nd Cir. 1997).

heroin, 1,964 grams for the seven trips with at least a 70 percent probability of carrying heroin, and 2,110 grams for all eight trips (the probability for the first trip was 65 percent).[69]

The appeals court found no fault with the district court's consideration of statistical models and probabilities in the sentencing hearing per se. Its continued objection was that "the statistical and economic analyses relate to drug trafficking generally and not to Shonubi specifically."

6.4.3 Sampling Acceptable

The appeals court in *Shonubi* distinguished between the use of probability-based extrapolation analyses for determination of the "relevant conduct" quantities carried by Shonubi on his seven prior trips and its use for estimating the quantity carried by Shonubi on his eighth trip.[70]

[Any] seeming inconsistency fails to take account of the different purposes for which the two estimates were made. The estimate of the quantity carried on the eighth trip was made to determine the quantity for the counts on which Shonubi was convicted. The estimate for the prior trips was used to punish Shonubi for conduct of which he had not even been charged, much less convicted. The distinction warrants caution in the use of estimates. Furthermore, the extrapolation as to the eighth trip was based on evidence of what Shonubi had done; the extrapolation for the prior seven trips was based on what 117 other people had done.[71]

Professor Finkelstein pointed out a further distinction, which he advanced in the context of distinguishing between extrapolation from four balloons to the 103 balloons carried by Shonubi on his eighth trip and extrapolation from the eighth trip to the prior seven trips. The first extrapolation involves a statistical sample, in which the mechanism for selection is randomization, while the second involves an observational study, in which the method of selection might be correlated with biasing factors, referred to as confounders. Professor Finkelstein's distinction applies with special force to extrapolation based on the 117 couriers reported by the DEA.[72]

---

[69] Ibid.

[70] We discussed this latter estimate in Chapter 4.

[71] Ibid. at 1092.

[72] Ibid. fn. 4.

6.5 Summary

The courts distinguish probability statistics offered by an expert concerning circumstantial evidence about which he or she states no conclusions, which are inadmissible, from probability statistics as a basis for an expert's conclusions concerning real evidence, which are admissible.

One should support forensic evidence — fingerprints, blood tests, and DNA tests — with expert testimony concerning facts and frequencies, not probabilities. For example, "Of blood samples taken from 4,500,234 individuals, all were distinguishable in at least one of 300 base pairs."

When a case is tried before a jury, do not attempt to combine separate pieces of evidence into a single all-embracing probability, at least not directly and not by the product rule.

When trying a case before a judge, statistical models, probability, and estimation methods may be introduced into evidence.

6.6 To Learn More

The *Collins* case generated a great deal of commentary when the decision first appeared; see, for example, Farley and Mosteller [1979] as well as the comments that appeared in the *Minnesota Law Review* and the *Duke Law Journal*. Kaye [1997] writes extensively about the use of DNA evidence. Monahan and Walker [1985] discuss the use of psychological profiles.

Chapter 7

Civil Law

A case should never be left to a jury simply on a question of probabilities with a direction to find in accordance with the greater probability. Probabilities may help out items of evidence from which an inference can be drawn, but cannot take their place. To allow a jury to dispose of a case simply upon a weighting of the probabilities is to turn them loose into the field of conjecture, and to have the rights of the parties determined by guess.[1]

Presumptions are indulged in to supply the place of facts. They are never allowed against ascertained and established facts. When these appear, presumptions disappear.[2]

In this chapter, we consider the application and limited acceptance of probability theory in civil hearings.

7.1 The Civil Paradigm

You go for a walk to get some fresh night air, only to wake an hour later with tire tracks on your face and testimony from eyewitnesses that a big

---

[1] *Virginia and Southwest R.R. Company v. Hawk*, 160 Fed 348 (1908), 352, 87 C.C.A. 300, 304.

[2] *Lincoln v. French*, 105 U.S. 614, 617 (1881).

yellow taxi knocked you down. The Yellow Cab company owns 80% of the town's taxicabs, so naturally you sue Yellow Cab.

You make it to Woodstock for the festival. What a story to tell your children. Leaving the grounds three days later, very tired, very wet and very hungry, you find yourself charged with fraud and trespass for sneaking in without a ticket. "I had a ticket," you assert. But 80% of the patrons didn't.

You live in Woburn, MA, and have been drinking the town's foul-tasting well water all your life. Now, you have leukemia. Your risk of getting leukemia, say the scientists, is five times what it would have been if you hadn't drunk the water. That is, 80% of the cases can be traced to those wells. Are the companies whose chemicals fouled the wells to blame for your illness in particular?

What is your opinion? Is it the same in all three cases? What about your colleagues? Are their reactions the same as yours? And, more important, what have the courts decided?

7.2   Holdings

For a Jew living in the Middle East in 200 A.D., the answer to the big yellow taxi could be found in the Talmud, though, as in modern courts, there was still room for both sides: "If nine shops sell ritually slaughtered meat and one sells meat that is not ritually slaughtered and he bought in one of them and does not know which one, it is prohibited because of the doubt; but if meat was found in the street, one goes after the majority."[3]

Modern civil courts seldom will accept arguments based on probabilities alone as demonstrated in the 1908 and 1881 cases quoted at the beginning of this chapter: "that an alleged fact is quantitatively probable is not probative evidence of its actual truth."[4]

Similar in its factual basis to the imaginary case of the big yellow taxi is *Guenther v. Armstrong Rubber Co.*[5] A defective tire caused the injury. The plaintiff could not be sure who manufactured the tire, but could demonstrate that 75 to 80% of the tires sold locally had been manufactured by Armstrong Rubber. The Third Circuit appeals court ruled the claim should not have gone to a jury as "the latter's verdict would be at best a guess."[6]

---

[3] Kethuboth 15a as quoted in Rabinovitch [1969].
[4] *Day v. Boston & Marine R.R.*, 96 Me. 207, 217-218, 52 A. 771, 774 (1902); see also *Toledo, St.L. & W.R. Co. v. How* 191 F. 776, 782-783 (6th Cir. 1911).
[5] 406 F.2d 1315 (3rd Cir. 1969).
[6] Id. at 1318.

In *Smith v. Rapid Transit Inc.*,[7] defendant was the only company that operated buses on the street where the accident occurred. "The most that can be said of the evidence in the instant case is that perhaps the mathematical chances somewhat favor the proposition that a bus of the defendant caused the accident. This is not enough."[8]

7.2.1 Exception for Joint Negligence

One exception to the preceding rule is that of joint negligence. In *Oliver v. Miles*,[9] two persons were hunting together and both shot across the highway. The plaintiff, traveling on the highway, was struck by pellets. Both parties were held liable as both were negligent. A similar shooting incident led to a similar result in *Summers v. Tice*.[10]

In *Hall v. E.I. DuPont De Nemours & Co., Inc.*,[11] the plaintiffs were children injured when blasting caps they were playing with exploded. Although the explosion destroyed the blasting caps, thus making it impossible to identify the manufacturer, the court ruled in favor of the plaintiffs. This case can be distinguished from *Guenther v. Armstrong Rubber Co.*[12] in that the defendants in *Hall* often acted as if they were a single entity. All failed to provide warning labels or take other preventive measures despite knowledge that children would and did play with blasting caps. And they jointly lobbied against legislation that would have required such precautions.

7.2.2 Exception for Expert Witnesses

The testimony of expert witnesses who speak in terms of probabilities may or may not be admissible, depending on the forum; if admissible, it may well be ignored.

A discussion of probabilities by a series of distinguished philosophers was admitted more than a century ago in the case of a disputed will,[13] only to be ignored by the judge in making his ruling.

A doctor's testimony that a plaintiff's death by cancer was likely to be causally related to his fall was unsupported by clinical evidence. As a

---

[7] 317 Mass. 469, 58 N.E.2d 754 (1975).
[8] Id. at 756. See also *Sawyer v. U.S.* 148 F. Supp. 877 (M.D. Ga. 1956).
[9] 144 Miss. 852, 110 So. 666 (1927).
[10] 33 Cal.2d 80 (1948).
[11] 345 F. Supp. 353 (E.D. N.Y. 1972).
[12] 406 F.2d 1315 (3rd Cir. 1969).
[13] *Robinson v. Mandell*, 20 F.Cas. 1027 (C.C.D. Mass. 1868); see also Meier and Zabell [1980].

mere mathematical likelihood, a "guesstimate" based on opinion alone, it was insufficient ruled a Massachusetts court.[14]

In a patent dispute over the highly successful anti-ulcer medication, Zantac®, the district court in *Glaxo, Inc. and Glaxo Group Ltd. v. Novopharm Ltd.*[15] found the X-ray diffraction evidence demonstrated "in clear and convincing fashion that Novopharm's product would not contain any of Form 2 RHCl and thus would not infringe the patents."[16] Glaxo argued that the only support for the district court's finding was the isolated statement of Glaxo's expert witness, Dr. Byrn: "I would put a probability of 60/40 that [an IR spectrum of Glaxo's Form 1 RHCl product is] not Form 2."

The appeals court dodged the issue as it ruled that statements by Novopharm's witness, Dr. Durig, as well as by Glaxo's own witnesses, Drs. Klinkert and Snyder, found credible by the district court, and not merely the testimony of Byrn, provided ample support for the court's factual finding that Glaxo failed to prove infringement under a single-peak analysis.[17]

7.2.3 Distinguishing Collins

> *People v. Collins* does not foreclose all use of statistical compilations as evidence to establish the existence of a fact. In *Collins*, the prosecution introduced evidence tending to establish that a robbery was committed by a couple in a yellow automobile, one, a blonde Caucasian woman wearing a pony tail, and the other, a Negro male with a beard and mustache. The prosecution assigned arbitrary numerical values to the probability of encountering those circumstances in a random population and, by applying a theory of mathematical probability known as the "product rule," purported to calculate with mathematical precision the probability of encountering the conjunction of those circumstances in a random population. In reversing defendant's conviction, the Supreme Court [of California] established restrictions on the assignment of numerical values to the probability of encountering a given set of circumstances in a random population, and the use to which such coefficients might be put.

---

[14] *King's Case*, 352 Mass 488 491-492, 225 N.E.2d 900,902 (1967).
[15] 110 F.3d 1562, 42 U.S.P.D.2d 1257 (4th Cir. 1997).
[16] Id. at 1566.
[17] Id. at 1567.

In the instant case[18] no attempt was made to assign a numerical figure to the probative value of the inferences which might be drawn from the statistical compilations. Moreover, the trier of fact was not asked to make a finding based on a mathematical theory of probability but merely to draw inferences from statistical facts derived from actual experience and observation. The various statistical compilations by the CHP, when considered as a whole, were reasonably reliable as indicators of appellant's level of efficiency as compared to those of officers performing like duties under like circumstances. (1) The comparisons were made over a sufficiently extended period of time so as to eliminate the effect of any fluctuations due to transitory conditions; (2) the individual and members of the group were performing comparable activities under comparable conditions; (3) the criteria used for comparison reflected the range of activities in which the individual and members of the group were engaged; and (4) the group with whom appellant was compared was sufficiently large and its members were selected on such a basis as would assure a fair representation of those performing like duties as appellant. The studies were competent and relevant on the issue of appellant's efficiency; their use was not subject to the strictures of *People v. Collins*.

7.2.4 Applying Bayes' Theorem

In the paternity suit involving the legendary comedian Charlie Chaplin, a jury was permitted to decide Chaplin fathered a child, although genetic testing demonstrated this was absolutely impossible.[19,20] Such a decision would be unlikely today.

In *Cole v. Cole*,[21] the court rejected the admission of an expert's testimony of a high probability of paternity derived via Bayes' formula because there was strong evidence the defendant was sterile as a result of a vasectomy.

> The source of much controversy is the statistical formula generally used to calculate the probability of paternity: the Bayes Theorem .... Briefly, the Bayes Theorem shows how new statistical information alters a previously established probability

---

[18] *Bodenschatz v. State Personnel Board*, 15 Cal. App. 3d 775, 781. (4th Dist. 1971).

[19] *Barry v. Chaplin*, 74 Cal. App. 2d 652 (1946).

[20] See also Berry and Geisser [1986], Berry [1991], and Ylvisaker [1986].

[21] 74 N.C. App. 247, aff'd., 314 N.C. 660 (1985).

.... When a laboratory uses the Bayes Theorem to calculate a probability of paternity it must first calculate a "prior probability of paternity" .... This prior probability usually has no connection to the case at hand. Sometimes it reflects the previous success of the laboratory at excluding false fathers. Traditionally, laboratories use the figure 50% which may or may not be appropriate in a given case.

Critics suggest that this prior probability should take into account the circumstances of the particular case. For example if the woman has accused three men of fathering her child or if there are reasons to doubt her credibility, or if there is evidence that the husband is infertile, as in the present case, then the prior probability should be reduced to less than 50%.[22]

The question remains as to what value to assign the prior probability, and whether absent sufficient knowledge to pin down the prior probability with any accuracy, we can make use of Bayes' theorem at all. At trial, an expert called by the prosecution in *Plemel v. Walter*[23] used Bayes' theorem to derive the probability of paternity.

If the paternity index or its equivalents are presented as the probability of paternity, this amounts to an unstated assumption of a prior probability of 50 percent. ... the paternity index will equal the probability of paternity only when the other evidence in this case establishes prior odds of paternity of exactly one.[24]

The expert is unqualified to state that any single figure is the accused's "probability of paternity." As noted above, such a statement requires an estimation of the strength of other evidence presented in the case (i.e., an estimation of the prior "probability of paternity"), an estimation that the expert is no better position to make than the trier of fact.[25]

Studies in Poland and New York City have suggested that this assumption [a 50 percent prior probability] favors the putative father because in an estimated 60 to 70 percent of paternity cases the mother's accusation of paternity is correct. Of course,

---

[22] Id. at 328.
[23] 303 Or. 262 (1987).
[24] Id. at 272.
[25] Id. at 275.

the purpose of paternity litigation is to determine whether the mother's accusation is correct and for that reason it would be both unfair and improper to apply the assumption in any particular case.[26]

A remedy was offered:

If the expert testifies to the defendant's paternity index or a substantially equivalent statistic, the expert must, if requested, calculate the probability that the defendant is the father by using more than a single assumption about the strength of the other evidence in the case .... If the expert uses various assumptions and makes these assumptions known, the fact finder's attention will be directed to the other evidence in the case, and it will not be misled into adopting the expert's assumption as to the correct weight to be assigned the other evidence. The expert should present calculations based on assumed prior probabilities of 0, 10, 20, ..., 90 and 100 percent.[27]

The courts of many other states have followed *Plemel*. "The better practice may be for the expert to testify to a range of prior probabilities, such as 10, 50 and 90 percent, and allow the trier of fact to determine which to use."[28]

7.3 Speculative Gains and Losses

[It] is now an accepted principle of contract law ... that recovery will be allowed where plaintiff has been denied an opportunity or chance to gain an award or profit even where damages are uncertain.[29]

In *Chaplin v. Hicks*,[30] an actress sued because she had been denied a tryout. Hicks, a theatre manager, had run a contest for actresses. Contestants, including Miss Chaplin, submitted photographs that were published subsequently in a number of newspapers. Readers were able to vote for

---

[26] Id. at 276, fn. 9.
[27] Id. at 279. See also Kaye [1988].
[28] *County of El Dorado v. Misura*, 33 Cal. App. 4th 73 (1995) citing *Plemel, supra*, at 1219; *Peterson* (1982 691, fn. 74), *Paternity of M.J.B.*, 144 Wis.2d 638, 643; *State v. Jackson*, 320 N.C. 452, 455 (1987), and *Kammer v. Young*, 73 Md. App. 565, 571 (1988). See also *State v. Spann*, 130 N.J. 484, 499 (1993).
[29] *Miller v. Allstate Ins. Co.*, 573 So. 2d 24, 29 (Fla. 1990).
[30] 2 King's Bench 706 (1911).

their favorites and Hicks promised tryouts to the 50 top vote getters. Chaplin was among the top 50, but Hicks failed to meet with her. Although the possibility that she would land a paying part subsequent to an interview was remote, the trial court awarded her £100 in compensation and the award was upheld on appeal.

For further applications of probability in contract law, see Eisenberg [1998]. For aid in determining compensation for lost wages, loss of services, and foregone profits, see Hall and Lazear[1994] after reading Chapters 11 and 12.

7.4 Summary

Modern civil courts seldom accept arguments based on probabilities alone. This is particularly true in the cases of alleged discrimination discussed in Chapter 12. Exceptions lie with contract law and in assessing future losses. In paternity cases, a range of prior probabilities may be considered in the context of Bayes' theorem. A further exception lies with epidemiological evidence, a topic we consider in the next chapter.

7.5 To Learn More

In the late 1960s and early 1970s, a number of authors tried to show how probability might be used in the criminal courtroom, among them Kaplan [1968], Cullison [1969], and Finkelstein and Farley [1970]. Tribe [1971][31] and Kaye [1989] provide a healthy skepticism. Still such polemics continue, e.g., Aitken [1995], Kaye and Koehler [1991], and Koehler [1991], but have garnered few if any jurists among their adherents.

---

[31] With an attempt at rebuttal by Finkelstein and Farley [1971].

Chapter 8

Environmental Hazards

Probability that one or more team members may be infected by intruder organism: 75%.

Number of hours until entire world is infected: 2800.

From John Carpenter's movie, *The Thing.*

Experience has shown that opposite opinions of persons professing to be experts may be obtained to any amount ... wearying the patience of both court and jury, and perplexing instead of elucidating the questions involved.[1]

8.1 Concepts

The paradigm of the big yellow taxi no longer holds when disease of environmental origin is a possibility and many different individuals, if not an entire population, are at risk. Here, the courts can and will take notice of probabilistic causation, of possibilities in place of absolutes.

In this chapter, we consider a number of issues affecting environmental law, some exclusively legal in nature, some purely statistical, the majority involving both. We study criteria for the admissibility and sufficiency of statistical (epidemiological) evidence and learn of the emphasis placed by

---

[1] *Winan v. New York & Erie R.R.*, 62 U.S. 88, 100 (1853).

the courts on the standardized mortality ratio. We learn that risk is composed of two elements: (1) the magnitude of the loss, and (2) the probability of the loss. In dealing with competing risks, the courts have clearly favored models with a causal basis over the purely statistical. Last, we study the courts' treatment of multiple defendants and their application of the "market-share" rule.

8.2   Is The Evidence Admissible?

Daubert v. Merrell Dow Pharmaceuticals, Inc.[2] enlarged and confirmed the gatekeeping role of the district court in appraising the admissibility of scientific evidence.

Prior to 1993, the standard had been *Frye v. U.S.*[3] which held that expert opinion based on a scientific technique is inadmissible unless the technique is "generally accepted" as reliable in the relevant scientific community. In *Daubert*, the Supreme Court held that Federal Rules of Evidence, not *Frye*, provide the standard for admitting expert scientific testimony in a federal trial.[4] Recently, the Supreme Court ruled that *Daubert* applies to technical or specialized expert testimony as well.[5]

8.2.1   Daubert

Jason Daubert and Eric Schuller were born with serious birth defects. They and their parents alleged the birth defects had been caused by their mothers' ingestion of Bendectin, a prescription anti-nausea drug. Merrell Dow, which marketed the drug, moved for summary judgment, contending that Bendectin did not cause birth defects in humans and that petitioners would be unable to come forward with any admissible evidence to the contrary.

> Merrell Dow submitted an affidavit of Steven H. Lamm, physician and epidemiologist, and a well-credentialed expert on the risks from exposure to various chemical substances. Dr. Lamm stated he had reviewed all the literature on Bendectin and human birth defects — more than 30 published studies involving over 130,000 patients, and that no study had found Bendectin to be a substance capable of causing malformations in fetuses.

---

[2] 509 U.S. 579 (1993).

[3] 54 App. D.C. 46, 293 F. 1013 (1923).

[4] *Frye* remains the standard in many state courts, e.g., Arizona, even today (Hamilton [1998]).

[5] *Kumho Tire Co. v. Carmichael,* 526 U.S. 137 (1999).

Petitioners responded with the testimony of eight experts of their own, each of whom also possessed impressive credentials. These experts concluded that Bendectin can cause birth defects, conclusions based upon both test tube and live animal studies that found a link between Bendectin and malformations; pharmacological studies of the chemical structure of Bendectin that purported to show similarities between the structure of the drug and that of other substance known to cause birth defects; and the reanalysis of previously published epidemiological (human statistical) studies.

The District Court granted respondent's motion for summary judgment. The court stated that scientific evidence is admissible only if the principle upon which it is based is "sufficiently established to have general acceptance in the field to which it belongs."[6] The court concluded that petitioners' evidence did not meet this standard. Given the vast body of epidemiological data concerning Bendectin, the court held that expert opinion that is not based on epidemiological evidence is not admissible to establish causation.[7] Thus, the animal cell studies, live animal studies, and chemical structure analyses on which petitioners had relied could not raise, by themselves, a reasonably disputable jury issue regarding causation.

Petitioners' epidemiological analyses, based as they were on recalculations of data in previously published studies that had found no causal link between the drug and birth defects, were ruled to be inadmissible because they had not been published or subjected to peer review.

The U.S. Court of Appeals for the Ninth Circuit affirmed,[8] stating that expert opinion based on a scientific technique is inadmissible unless the technique is "generally accepted" as reliable in the relevant scientific community;[9] expert opinion based on a methodology that diverges significantly from the procedures accepted by recognized authorities in the field cannot be shown to be "generally accepted as a reliable technique."[10]

---

[6] 727 F. Supp. 570, 572 (S.D. Cal. 1989), quoting *U.S. v. Kilgus*, 571 F.2d 508, 510 (CA9 1978).

[7] 727 F. Supp. at 575.

[8] 951 F.2d 1128 (1991), citing *Frye v. U.S.*, 54 App. D.C. 46, 47, 293 F. 1013, 1014 (1923).

[9] 951 F.2d at 1129–1130.

[10] Id. at 1130, quoting *U.S. v. Solomon*, 753 F.2d 1522, 1526 (CA9 1985).

The court emphasized that other Courts of Appeals considering the risks of Bendectin had refused to admit reanalyses of epidemiological studies that had been neither published nor subjected to peer review.[11] Those courts had found unpublished reanalyses "particularly problematic in light of the massive weight of the original published studies supporting [respondent's] position, all of which had undergone full scrutiny from the scientific community."[12] Contending that reanalysis is generally accepted by the scientific community only when it is subjected to verification and scrutiny by others in the field, the Court of Appeals rejected petitioners' reanalyses as "unpublished, not subjected to the normal peer review process, and generated solely for use in litigation."[13] The court concluded that the petitioners' evidence provides an insufficient foundation to allow admission of expert testimony that Bendectin caused their injuries and, accordingly, that petitioners could not satisfy their burden of proving causation at trial.

8.2.2   Role of the Trial Judge

The Supreme Court overturned this ruling.[14] In Justice Blackmun's words, "'general acceptance' is not a necessary precondition to the admissibility of scientific evidence under the Federal Rules of Evidence, but the Rules of Evidence — especially Rule 702 — do assign to the trial judge the task of ensuring that an expert's testimony both rests on a reliable foundation and is relevant to the task at hand. Pertinent evidence based on scientifically valid principles will satisfy those demands."[15]

Recently, Justice Scalia wrote,[16]

> I join the opinion of the Court, which makes clear that the discretion it endorses — trial–court discretion in choosing the manner of testing expert reliability — is not discretion to abandon the gatekeeping function. I think it worth adding that it is not discretion to perform the function inadequately. Rather,

---

[11] 951 F.2d at 1130–1131.

[12] Id. at 1130.

[13] Id. at 1131.

[14] 509 U.S. 579 (1993).

[15] Id. at 597.

[16] *Kumho Tire Co. v. Carmichael*, 526 U.S. 137 (1999). The principal opinion in this case also makes clear that *Daubert* applies to all expert witnesses, whether scientist or not; presumably this includes statisticians.

it is discretion to choose among reasonable means of excluding expertise that is false and science that is junky. Though, as the Court makes clear today, the *Daubert* factors are not holy writ; in a particular case the failure to apply one or another of them may be unreasonable, and hence an abuse of discretion.

In Federal courts today, cross-examination, presentation of contrary evidence, and careful instruction on the burden of proof, rather than wholesale exclusion under an uncompromising general acceptance standard, are the appropriate means by which evidence based on valid principles may be challenged.

Pretrial Report

Federal Rule of Civil Procedure 26 requires the proposed expert witness to submit a written statement prior to trial detailing the following:

- Statement of opinions
- Basis of opinions
- Tables and graphs supporting the opinions
- Statement of qualifications
- List of all cases in the preceding four years in which he or she testified

---

8.3   Is the Evidence Sufficient?

8.3.1   SMR Defined

Epidemiology is the study of disease patterns in human populations, an attempt "to define a relationship between a disease and a factor suspected of causing it."[17] Epidemiological evidence is not only admissible in toxic and carcinogenic tort actions, it is indispensable where direct proof of causation is lacking.

---

[17] *Brock v. Merrell Dow Pharmaceuticals, Inc.*, 874 F.2d 307, 311 (5th Cir.), modified on reh'g, 884 F.2d 166 (5th Cir. 1989), cert. denied, 494 U.S. 1046 (1990).

Epidemiologists speak in the statistical language of risks and probabilities. The relative risk that exposure to a given causal factor (c) will lead to a certain disease (d) is expressed as a single-digit ratio, known as the "standardized mortality ratio" (SMR). An SMR of 1.0 is the expected rate of contracting d in a population not influenced by c, the causal factor under investigation. An SMR of 2.0 means that d was as likely as not to have been caused by c, and an SMR greater than 2.0 means that d was more likely than not caused by c.[18]

In order to present a jury question on the issue of causation, you must demonstrate that some factor c is "more likely than not" the cause of a disease d. This burden can be met either through studies conclusively establishing a standardized mortality ratio of more than 2.0, or through epidemiological evidence falling short of 2.0 in combination with "clinical or experimental evidence which eliminates confounding factors and strengthens the connection between c and d specifically in the circumstances surrounding the plaintiff's case of d."[19]

Checklist for Admissibility of Scientific Evidence

- Has the theory or technique been tested or can it be tested?
- Has it been subjected to peer review and publication?
- What are the potential sources of error?
- If a technique, what standards control its operation?
- Has it gained widespread acceptance within a relevant scientific community?

In a follow-up to *Daubert,* the Ninth Circuit ruled that "plaintiffs must establish not just that Bendectin increased somewhat the likelihood of birth defects, but that it more than doubled it — only then can it be said that Bendectin is more likely than not the cause of their injury."[20]

---

[18] *In Re Joint Eastern & Southern District Asbestos Litigation*, 52 F.3d 1124, 1128 (2nd Cir. 1995), quoting *Manko v. U.S.*, 636 F. Supp. 1419, 1434 (W.D. Mo. 1986), aff'd and remanded, 830 F.2d 831 (8th Cir. 1987).
[19] Ibid.
[20] *Daubert v. Merrell Dow Pharmaceuticals, Inc.*, 4 F.3d 1311, 1320 (9th Cir. 1995).

An alternate viewpoint, that insisting that the SMR exceed 2.0 fails to achieve the desired policy goals, is presented by Parascandola [1998].

8.3.2 Sufficiency Defined

In Re Joint Eastern & Southern District Asbestos Litigation[21] marked the convergence of epidemiological evidence, probabilistic causation in carcinogenic torts, and the important issue of the extent to which a trial court might assess the sufficiency of scientific evidence. In the court's words, "The central question is the standard governing federal judges' evaluations of the *sufficiency* — as opposed to admissibility — of scientific evidence already admitted."[22]

The scientific community is divided on whether asbestos exposure significantly increases the risk of contracting colon cancer. At trial in the U.S. District Court for the Southern District of New York (Robert W. Sweet, Judge), both plaintiff Arlene Maiorana and one of the defendants, U.S. Mineral Products Company (USMP), brought expert witnesses and numerous epidemiological studies to bear on their dispute over the causal link between asbestos and colon cancer.

In a series of rulings in 1991, the district court awarded summary judgment in favor of defendants, including USMP, on the grounds that the epidemiological and clinical evidence of causation was insufficient to meet the preponderance standard.[23]

The federal appeals court reversed the grant of summary judgment and remanded the case for further proceedings, concluding that the evidence was sufficient to survive summary judgment.[24] The appeals court found that the plaintiff had presented not only epidemiological studies in support of a causal connection between asbestos exposure and colon cancer, but also clinical evidence in the form of Maiorana's own medical records and personal history, which plaintiff's experts used to exclude other possible causal factors. The statements of the plaintiff's experts, viewed in the light most favorable to plaintiff, were the "equivalent of stating that asbestos exposure more probably than not caused the colon cancer."[25]

---

[21] 52 F.3d 1124 (2nd Cir. 1995).

[22] Id. at 1126.

[23] See *In Re Joint Eastern & Southern District Asbestos Litigation*, 758 F. Supp. 199 (S.D. N.Y. 1991) (Asbestos Litigation I), reargument denied, 774 F. Supp. 113, reconsideration denied, 774 F. Supp. 116 (1991).

[24] *In Re Joint Eastern & Southern District Asbestos Litigation*, 964 F.2d 92, 96–97 (2nd Cir. 1992) (Asbestos Litigation II).

[25] Id. at 97.

The district court granted USMP's motion for judgment as a matter of law, basing its decision on its findings that (1) plaintiff's epidemiological evidence was insufficient to support a causal connection between asbestos exposure and colon cancer, and (2) plaintiff had failed to present affirmative clinical evidence to overcome the paucity of statistically significant epidemiological proof.[26]

> In order for plaintiff to present a jury question on the issue of causation, the district court noted that she bore the burden of demonstrating that asbestos exposure was "more likely than not" the cause of her husband's colon cancer.[27] This burden could be met either through studies conclusively establishing an SMR of more than 2.0, or through epidemiological evidence falling short of 2.0 in combination with "clinical or experimental evidence which eliminates confounding factors and strengthens the connection between c and d specifically in the circumstances surrounding the plaintiff's case of d."[28]

The district court then evaluated the epidemiological studies in evidence, noting the SMRs reached in each study and assessing the reliability of each study in light of five sufficiency criteria derived from the work of Hill [1971]:

1. Strength and consistency of association between c and d
2. Dose-response relationship to c of d
3. Experimental evidence
4. Plausibility
5. Coherence

The meanings of each of these sufficiency criteria will be described in turn as we reproduce at length portions of Judge Sweet's opinion.

8.3.3 Strength and Consistency of Association

Strength of association is "measured by the relative risk or the ratio of the disease rate in those with the factor to the rate in those without."[29]

---

[26] *In Re Joint Eastern & Southern District Asbestos Litigation*, 827 F. Supp. 1014, 1050–1051 (S.D. N.Y. 1993) (Asbestos Litigation III).

[27] Id. at 1029–1030, citing *In Re Agent Orange Product Liability Litigation*, 597 F. Supp. 740, 785 (E.D. N.Y. 1984) (Agent Orange I), aff'd, 818 F.2d 145 (2nd Cir. 1987), cert. denied, 484 U.S. 1004 (1988).

[28] Id. at 1030.

[29] Id. at 1038, quoting *Landrigan v. Celotex Corp.*, 127 N.J. 404, 605 A.2d 1079, 1086 (1992).

Consistency of association is "measured by comparing the association between a purported cause and effect identified in one study with the results of other studies and with other relevant scientific knowledge."

Plaintiff's expert witnesses testified at trial that there was a causal relationship between asbestos exposure and colon cancer. The district court conducted an independent and detailed analysis of many of the epidemiological studies and concluded that there was no basis for plaintiff's experts' conclusions. To this end, the district court noted that many of the studies yielded SMRs falling between 1.0 and 1.5, which the court deemed "statistically insignificant."[30]

A few studies adduced by plaintiff did yield SMRs that the court acknowledged were statistically significant, SMRs of 1.62, 1.85, and 2.27.[31] The district court criticized the methodologies employed in these studies and found that when considered in the context of all the studies, plaintiff's evidence "establishe[d] only the conclusions that the association between exposure to asbestos and developing colon cancer is, at best, weak, and that the consistency of this purported association across the studies is, at best, poor."[32] The district court concluded that the epidemiological evidence failed to satisfy the Sufficiency Criterion of strength and consistency of association, and therefore failed to contribute to the sufficiency of plaintiff's proof on causation.[33]

8.3.4 Dose-Response Relationship

The district court described the dose-response relationship, the relationship between certain doses of c and the subsequent development of d, between asbestos and colon cancer as "erratic at best."[34] In his analysis, the district judge refused to credit a study cited by plaintiff suggesting a relatively high dose-response relationship, because the court disagreed with the assumption of that study that lung cancer rates could be used in the absence of any direct measurement as a substitute measure for a given population's exposure to asbestos.[35]

---

[30] Id. at 1041–1042.
[31] Id. at 1040–1041.
[32] Id. at 1042.
[33] Id. at 1043.
[34] Id. at 1044.
[35] Id. at 1044–1045.

8.3.5 Experimental Evidence

The district court found that studies done on animals also have not established a causal relationship between asbestos and colon cancer.[36] Although plaintiff's expert testified that asbestos exposure did increase the rate of development of pre-cancerous polyps in animals, the district court without directly discussing the studies cited by him concluded that the experimental evidence lent no support to the claim that asbestos exposure and colon cancer were causally related.[37]

8.3.6 Plausibility

This sufficiency criterion asks whether it is biologically plausible, in light of the biological and chemical mechanisms involved, for exposure to c to precipitate the subsequent development of disease d.[38]

The district court acknowledged that asbestos is generally a carcinogen, and that a causal linkage between asbestos exposure and colon cancer is "possible." Nonetheless, the district court without citing any specifics stated that plaintiff's evidence "does not support the conclusion that [the] relationship [between asbestos and colon cancer] is anything more than possible.[39]

8.3.7 Coherence

This criterion refers to the analysis of the instant causal factor in the context of other possible causal factors. For example, a worker's exposure to asbestos as a possible cause of his lung cancer would be lessened on a showing that he was a heavy smoker. In meeting the coherence criterion, then, plaintiff would have to use epidemiological evidence to rule out other possible confounding factors, or she would have to employ clinical or other particularistic evidence to eliminate the possibility that other confounding factors were more likely than not to have caused the disease.[40]

The district court first observed that not only were plaintiff's experts unable to narrow down the universe of possible confounding factors for

---

[36] Id. at 1046.
[37] Id. at 1045–1046.
[38] Id. at 1038.
[39] Id. at 1046.
[40] Id. at 1046; see also the discussion of multiple regression in Chapter 12.

colon cancer, but they acknowledged that asbestos exposure is not considered to be a risk factor.[41] Indeed, as the district court noted, the rate for colon cancer in Nassau County, the New York county in which Maiorana resided, was 25.7 per 100,000 persons between 1970 and 1975, compared to 18.1 per 100,000 persons nationally during the same time period. This fact, the district court wrote, "raises a serious question about the various carcinogenic substances to which Maiorana and all the other residents of Nassau County have been exposed over the years and which constitute confounding factors in assessing the proximate cause of Maiorana's cancer."[42]

With respect to clinical evidence, the district court observed that plaintiff's case consisted only of a "differential" diagnosis, meaning that plaintiff's efforts attempted to exclude other confounding factors based on Maiorana's medical records as opposed to introducing affirmative evidence of causation such as the presence of asbestos fibers in Maiorana's cancerous tissues. Plaintiff's experts testified that because Maiorana was only 40 years old at the time of his death, had no family history of cancer, suffered from no special disease or syndrome, and did not face an abnormal risk in his diet inasmuch as it was low in fat, his colon cancer must have been caused by asbestos exposure.[43] The district court concluded, however, that this expert testimony "failed to contribute to the sufficiency of the Plaintiff's causation proof." Although the district judge acknowledged that the presence of colon cancer in a 40-year-old man was uncommon, he nevertheless concluded that "it is neither startling nor so uncommon that it constitutes a mesothelioma-like signature disease arising only when a person of that age is exposed to asbestos." The district court further pointed out that plaintiff's latency period from his first exposure to asbestos was only 13 years, a time period that the court found was too short to be probative, in light of studies revealing a typical latency period of more than 20 years from the initial exposure to asbestos.[44]

In sum, the district court found that inasmuch as plaintiff's epidemiological evidence failed, in the court's view, to satisfy any of the Sufficiency Criteria commonly employed by epidemiologists, this evidence was insufficient to support the general proposition that asbestos exposure causes colon cancer.[45] The district court criticized plaintiff's experts' conclusions as "masquerading behind the guise of sound science." Given plaintiff's

---

[41] Id. at 1047.

[42] Id. at 1048.

[43] Id. at 1049.

[44] Id. at 1049–1050.

[45] Id. at 1050.

failure to show that any asbestos fibers were found in Maiorana's cancerous tissues, the district court found that the sum total of plaintiff's evidence did not justify the jury's finding of causation, and that "the jury's finding could only have been the result of sheer surmise and conjecture."[46] Therefore, concluded the district court, entry of judgment as a matter of law in favor of defendants was warranted.

8.3.8 Other Discussions of Sufficiency

As illustrations of appropriate uses of summary judgment or judgment as a matter of law, the *Daubert* court cited two courts of appeal opinions mandating, on grounds of insufficiency, the removal from the jury of cases grounded in epidemiological evidence. In the first case, *Turpin v. Merrell Dow Pharmaceuticals, Inc.*,[47] the Sixth Circuit found that experimental evidence involving animal studies was insufficient to permit a reasonable jury to find that Bendectin, an anti-nausea prescription for morning sickness, caused plaintiff's birth defects. Unlike Arlene Maiorana, the plaintiffs in *Turpin* did not present supportive epidemiological evidence in addition to their animal studies, whereas their adversaries introduced 35 epidemiological studies demonstrating no causal link.[48] "We find nothing in *Turpin* that suggests an alteration of the traditional standard for assessing the sufficiency of scientific evidence."[49]

In the second case, *Brock v. Merrell Dow Pharmaceuticals, Inc.*,[50] the Fifth Circuit found that plaintiff's epidemiological evidence was insufficient to support a jury verdict finding that Bendectin caused plaintiff's child's birth defects; thus the court granted judgment as a matter of law in favor of defendants. The basis for the court's holding was that the risk ratios yielded by the plaintiff's epidemiological studies upon adjustment by confidence intervals included the ratio of 1.0, which would be the standard expected rate of birth defects in a population not exposed to Bendectin ingestion.[51]

---

[46] Id. at 1051, quoting *Samuels v. Air Transport Local 504*, 992 F.2d 12, 14 (2nd Cir. 1993).

[47] 959 F.2d 1349 (6th Cir.), <u>cert. denied</u>, 113 S. Ct. 84 (1992).

[48] Id. at 1353.

[49] 52 F.3d at 1133.

[50] 874 F.2d 307, 311 (5th Cir.), <u>modified on reh'g</u>, 884 F.2d 166 (5th Cir. 1989), <u>cert. denied</u>, 494 U.S. 1046 (1990).

[51] 52 F.3d at 1133, quoting *Brock*, 874 F.2d at 312–313.

Our reading of *Brock* is that it applied the traditional sufficiency standard.[52]

Although we conclude that *Daubert* did not alter the traditional sufficiency standard contrary to the position urged by USMP we acknowledge that sufficiency poses unique difficulties for trial courts in toxic or carcinogenic tort cases, such as the one before us, which hinge on competing interpretations of epidemiological evidence. By its nature, epidemiology is ill suited to lead a factfinder toward definitive answers, dealing as it does in statistical probabilities and the continual possibility of confounding causal factors. "One difficulty with epidemiological studies is that often several factors can cause the same disease."[53] In light of the inherent uncertainty shrouding issues of probabilistic causation, the decision of a district court on whether plaintiff's epidemiological evidence is sufficient to get to the jury should be guided by the well-established standards governing judgment as a matter of law whether, viewed in the light most favorable to the nonmoving party, "the evidence is such that, without weighing the credibility of the witnesses or otherwise considering the weight of the evidence, there can be but one conclusion as the verdict that reasonable [jurors] could have reached."[54]

Applied to epidemiological studies, the question is not whether there is some dispute about the validity or force of a given study, but rather, whether it would be unreasonable for a rational jury to rely on that study to find causation by a preponderance of the evidence. In addition, multiple epidemiological studies cannot be evaluated in isolation from each other. Unlike admissibility assessments, which involve decisions about individual pieces of evidence, sufficiency assessments entail a review of the sum total of a plaintiff's evidence.[55]

---

[52] Ibid.

[53] See *Brock*, 874 F.2d 311.

[54] 52 F.3d at 1133, quoting *Samuels v. Air Transport Local 504*, 992 F.2d 12 at 14 (2nd Cir. 1993), quoting *Simblest v. Maynard*, 427 F.2d 1, 4 (2nd Cir. 1970).

[55] Ibid.

8.4   Risk versus Probability

Mathematical probability is not the ultimate test of foreseeability, duty, or negligence. Probability of injury is only one of many factors to be evaluated in the duty/breach of duty analysis and is a factor of varying significance. "If the risk is an appreciable one, and the possible consequences are serious, the question is not one of mathematical probabilities alone. The odds may be 1000 to 1 that no train will arrive at the very moment that an automobile is crossing a railway track, but the risk of death is nevertheless sufficiently serious to require the driver to look for the train …. As the gravity of the possible harm increases, the apparent likelihood of its occurrence need be correspondingly less."[56]

Consider how ineffective the anti-littering ordinances are in those cities where the fine for littering is a mere $50. Most potential litterbugs are willing to take their chances that a police officer will not drive by just as they toss their garbage out the window. In Del Mar, CA, the fine is $1000 for the first offense. Makes you think, doesn't it?

In *Allen et al. v. U.S.*,[57] a large number of plaintiffs sued on behalf of themselves and their minor children, alleging that radioactive fallout from nuclear tests caused a variety of cancers and leukemia. The principal issue was whether the U.S. government was negligent in performing these tests and in failing to inform residents in the likely fallout areas of the probable consequences. A second major issue was whether the fallout was a substantial factor in causing the cancers.

Although the government denied all the plaintiff's allegations and moved for a summary dismissal of the case, it was soon brought out at trial that the government knew that radiation caused cancer, especially childhood leukemia. It was also shown that the government knew there would be radioactive fallout, and relied more on hope than on any concrete expectations that the fallout would not affect heavily populated areas. To compound the problem, the government had engaged in a campaign of disinformation designed to ensure local residents that radiation was somehow good for them.[58]

As is so often the case with epidemiological evidence, no one could conclude with certainty that the radiation from these tests caused the specific tumors in question. Still, testimony prior to trial revealed that someone standing outside in St. George, Utah a few minutes after the

---

[56] Prosser, *Handbook of the Law of Torts*, §32 at 147 (4th ed. 1971).

[57] 588 F. Supp. 247 (1984).

[58] If this seems hard to believe, one need only consult back issues of newspapers published at that time.

fallout arrived would have received more exposure in an hour than atomic workers were permitted in a week.[59] The federal government at the time knew a number of precautions, including wearing hats, as well as showering and shampooing thoroughly after exposure, but no one thought to inform the public about them.

The federal government argued that the low *average* dose could not have hurt anyone, but statisticians know there can be major deviations from a so-called "average." When particulate matter is distributed over a wide area, it follows the Poisson distribution.[60] Suppose we were to divide the St. George area into a grid, with each square on the grid just about the size of a single individual walking from store to store or standing in the open and talking to someone else. If each square received an average of one dose, a sixth of the squares would receive two doses, an eighteenth would receive three doses, about a third of 1% would receive five doses, and so forth. The court in *Allen* took judicial notice that a low average exposure did not mean that some individuals might not have been exposed to higher maximal levels.

The government also argued that too much time had elapsed between the exposure to the fallout and the appearance of the tumors. The court took judicial notice that cancer has a latency period — it takes 20 cell generations for a single mutated cell to divide to form a tumor one centimeter in volume, just at the borderline of detectable size.

In the end, the arguments that weighed most heavily on the court were qualitative, factual, cause-and-effect descriptions that were biological, physical, and statistical in nature. While many plaintiffs' claims were excluded, such as injuries that could not have been radiation related or that had some other known cause, the court rejected the government's arguments and ruled that recovery of damages would be available to any plaintiffs who could establish:

1. They or the decedent resided in the area affected by the government's acts.
2. They or the decedent were exposed to fallout radiation in excess of background rates (that is, were outdoors during the critical period).
3. Their injuries or the decedent's were of a type consistent with those known to be caused by ionizing radiation.

---

[59] That is, a week of what was then thought to be an acceptable level of exposure; by today's better understood criteria, it would be more radiation than one ought to be exposed to over a two-year period.

[60] We discuss the Poisson and other distributions in Chapter 9.

8.4.1 Competing Risks

A heavy smoker developed lung cancer after 32 years of fighting fires, and 42 years of smoking cigarettes in *McAllister v. Workmen's Compensation Appeals Board*.[61] Nonetheless, the court found sufficient factual connection to keep the fireman's employer in the case, citing *Bethlehem Steel Co. v. Industrial Accident Comm*.[62] and stating that it is enough that "the employee's risk of contracting the disease by virtue of the employment must be materially greater than that of the general public."

8.5 Use of Models

I was part of the Motrin® development team, a drug you may be familiar with as Advil or ibuprofen. In ensuring this drug would meet FDA standards for safety and efficacy, I designed *controlled*[63] experiments on mice, dogs, and humans, *double blinded*[64] experiments guaranteed to pass review by the FDA.

One seldom has the opportunity in the courtroom to assess the results of carefully planned experiments. One must make do with the data at hand, and hope not merely to discern a pattern in them, but to be able to prove that the pattern and not simple chance is responsible for what is observed.

A personal injury attorney could probably make a good living today simply by perusing the morbidity reports issued by the Centers for Disease Control. The idea would be to locate in each state the county with the highest incidence of leukemia or some other terrifying disease, look for a polluter resident in that county — they're not hard to find — and then sue. Never mind that this is what is known as an ad hoc–post hoc hypothesis[65] or that some county is bound to have the highest incidence only because this is how "highest" is defined.

Absent the careful attention to detail of the controlled experiment, how can we be sure whether a specific factor (the polluter) is the true cause of the malady? Freedman [1999] proposes three criteria that must be satisfied:

1. Multiple studies of multiple groups by multiple observers all point in the same direction. For example, if the incidence of leukemia is high in a given area, it should be high for both men and women.

---

[61] 69 Cal. 2d 408, 71 Cal. Rptr. 697 (1968).
[62] 21 Cal. 2d 742, 8 Compensation Cases 61 (1943).
[63] See Section 13.2.
[64] See Section 13.2.4.
[65] See Section 13.6.

2. The relationship continues to hold even after the effects of potential confounding variables are taken into account by appropriate statistical techniques.
3. There is a plausible cause and effect explanation for the relationship.

This last criterion, first raised in Section 8.3.6, is perhaps the most compelling for both statisticians and jurists. Courts are more interested in commonsense arguments than they are in complex formulae, even though common sense may entail a biologist's or meteorologist's expert knowledge of cause and effect.

Shortly after the U.S. government launched a swine flu immunization program in 1976, a marked increase in reports of Guillain-Barre syndrome (GBS) caused it to halt distribution of the vaccine. Aware of a strong statistical correlation between swine flu vaccination and GBS in the first ten weeks after vaccination, it stipulated to liability and settled a variety of claims.

The plaintiffs in *Cook et al. v. U.S.*[66] all suffered onset of GBS from 12.5 to 13.5 weeks after vaccination, and the government denied responsibility because of the time lapse. Both sides relied on data collected by the Centers for Disease Control in Atlanta, but the two sides interpreted the data differently.

The first expert who testified for the government excluded from the raw data case reports from states whose reporting methods he considered unreliable. He calculated the base or background rate for GBS as 0.22 million cases per million persons per week. He demonstrated that among vaccinees this rate rose to 1 case per million per week in the first week, 2.5 in the second, peaking at 3.5 in the third, dropping back to 1 case in the fourth week, and trailing off gradually to the base rate well before the tenth week.

The first of the plaintiff's expert witnesses arrived at a somewhat different figure for the base rate. More important, he felt, was the sharp decline in case reports after December 16 when the decision to discontinue the vaccination program was made. The result, he claimed, was an underreporting of vaccine-related cases of GBS in subsequent weeks. For example, during the 12th and 13th weeks of the surveillance, the raw attack rates dropped from double-digit to single-digit figures, from about 1.7 to 0.6 per million. (This figure includes all individuals who had been inoculated from 1 to 12 weeks earlier). By adjusting for the hypothetical unreported cases, the expert arrived at a figure of 0.79 cases per million during the period from 13 to 18 weeks after vaccination, that is, more than twice the base rate.

---

[66] 545 F. Supp. 306 (N.D. Cal 1982).

Four experts testified in all, differing from one another in the following five respects:

- Whether to exclude case reports from states whose reporting methods were considered unreliable
- How to calculate the base or background rate for GBS — the expert's guesstimates varied from 0.22 to 0.44 million cases per million persons per week
- Extent of the underreporting that occurred once the decision to discontinue the vaccination program was made
- Whether the underreporting affected vaccinated and unvaccinated individuals equally
- Method used to correct for the underreporting

These disagreements served mainly to increase the court's distrust of statisticians. Pivotal to the court's ruling against the plaintiffs was its sense that the plaintiffs simply had not proved their case. A straightforward biological explanation could be provided for the initial rapid rise and fall of reported symptoms, as this is a characteristic of all viral infections. No similar biological explanation was offered for the alleged secondary rise. And, of course, the experts continued to disagree over whether such a rise actually occurred.

8.6 Multiple Defendants

Suppose the evidence yields an SMR on the order of 10 as in *A Civil Action*, but multiple sources of pollution are involved. If we argue as in our analogy of the yellow taxi in Section 7.1 that we cannot convict on probabilities alone, then an obvious polluter goes free; worse, future offenders are given a green light and the law fails to rein in clear-cut menaces to public health. *A Civil Action* (in which the plaintiffs lost) has proved the exception, not the rule.

In *Basko v. Sterling Drug, Inc.*, the plaintiff was blinded as a side effect of one or both of two drugs administered as treatment for a skin disease. The Court held "either force can be said to be the cause in fact of the harm, despite the fact the same harm would have resulted from either cause alone."[67]

---

[67] 416 F.2d 417 (2nd Cir. 1969).

Preventive Statistics

Epidemiological studies can be attacked on a variety of grounds including confounding; selection, response, and observer bias; and a change in classification methodology.

Confounding. Another variable, one correlated with the supposed hazard, may actually be responsible for the disease. Do storks bring babies? No, but in Europe one often finds many babies where there are many storks. Of course, this may be because European storks nest in rooftops, and where there are roofs, there are bound to be many new families underneath.

Bias.

1. *Selection.* Why do some people volunteer to participate in studies and others refuse to be counted? Those with grievances are far more likely to speak up than those who are busy and productive and have many other things to do.
2. *Response.* "Did you ever smoke marijuana?" For a study on AIDs, "How often do you have anal intercourse?" Can we really count on honest answers to questions?
3. *Observer.* Can you rely on your expert witness? Experts are no different from anyone else; they see what they want to see, or, worse, they see what they think you want them to see.

Hawthorne effect. Being part of an ongoing study causes subjects to modify their behavior.

Changes in Classification and Methodology. Changes in the reported incidence of a disease can be the result of (1) improvement in reporting methods, (2) changes in the disease classification (mental disease classification seems to undergo a minor revolution every ten years or so), (3) more accurate diagnoses as a disease is better understood, and (4) less accurate diagnoses as a disease grows more popular (e.g., Alzheimer's disease).

In *Sindell v. Abbott Laboratories*,[68] the plaintiff alleged injury due to cancer resulting from her mother's ingestion of diethylstilbestrol (DES) during pregnancy. The plaintiff could not identify the source of the DES. The *Sindell* court held all companies who manufactured DES at the time her mother took it to be jointly liable, and held each firm liable to the extent of its market share.

8.7 Summary

Federal Rules of Evidence — especially Rule 702 — assign to the trial judge the task of ensuring that an expert's testimony both rests on a reliable foundation and is relevant to the task at hand. The trial judge may find animal cell (*in vitro*) studies, live animal studies, computer simulations, and chemical structure analyses, as well as epidemiological analyses, that have not been published or subjected to peer review admissible.

Epidemiological evidence is not only admissible in toxic and carcinogenic tort actions, it is indispensable where direct proof of causation is lacking. In order to present a jury question on the issue of causation, you must demonstrate that some factor c is "more likely than not" the cause of a disease d. This burden can be met either through studies conclusively establishing a standardized mortality ratio (SMR) of more than 2.0, or through epidemiological evidence falling short of 2.0 in combination with "clinical or experimental evidence which eliminates confounding factors and strengthens the connection between c and d specifically in the circumstances surrounding the plaintiff's case of d."

The evidence must demonstrate (1) strength and consistency of association between c and d, (2) a dose-response relationship to c of d, (3) a risk of d by virtue of c that is materially greater than that of the general public, (4) plausibility, and (5) coherence.

The greater the potential harm, the greater the care that must be exercised, no matter how minute the probability of such harm may appear to be. It is the risk of the event, the product of the potential loss times its probability, that must be taken into consideration.

Scientific plausibility outweighs abstract statistical arguments.

Multiple potential plaintiffs may be held jointly liable. Liability may be determined by market share.

---

[68] 26. Cal.3d 588, 163 Cal. Rptr. 132, 607 P.2d 924 (1980).

HYPOTHESIS TESTING AND ESTIMATION

Ideally, one would provide the court with a survey of the entire population of interest. As we saw in Chapter 3, even the census falls short of this ideal so that, absent an unlimited amount of funds, we must content ourselves with a sample for most practical applications. Fortunately, if the sample is a representative one, free from bias, we may use it to estimate population characteristics and to test hypotheses concerning the population itself.

We also may provide estimates of the precision of our estimates, secure in the knowledge that as the sample grows larger, its properties grow ever closer to those of the population. But how large is large?

We cannot consider sample size alone; every statistical test involves three factors:

1. Sample size or sizes, the topic of the next chapter.
2. Significance level, the probability that chance alone may be responsible for a statistic's extreme value, considered in Chapter 10.
3. Power of a test, the probability of detecting a phenomenon of interest. Lack of power is an effective way to rebut poor statistical methodology, a topic we consider in Chapter 13.

Chapter 9

How Large Is Large?

Statistical evidence derived from an extremely small universe, as in the present case, has little predictive value and may be discarded.[1]

The officials lack contact with the common people; they're well prepared for the normal average trial, which rolls along its course almost on its own and needs only a push now and then, but faced with very simple cases or with particularly complex ones, they're often at a loss; because they're constantly constricted by the Law both night and day, they have no proper understanding of statistical relationships, and in such cases they feel that lack keenly.[2]

We know that as a sample grows larger, its characteristics more and more resemble the population from which it is derived. But how large is large? Federal agency guidelines for the establishment of statistical proof in discrimination cases require a showing that the protected group is selected at less than four-fifths or 80% of the rate achieved by the highest scoring group, but other factors may and often do intervene. In this chapter, we

---

[1] *Harper v. Trans World Airlines, Inc.*, 525 F.2d 409, 412 (8th Cir. 1975); accord, *Robinson v. City of Dallas*, 514 F.2d 1271, 1273 (5th Cir. 1975), *Morita v. Southern California Pemanente Medical Group*, 541 F.2d 217, 220 (9th Cir. 1976), <u>cert. denied</u>, 429 U.S. 1050 (1977).

[2] Paraphrased from Kafka's *The Trial*.

consider how the courts have answered this question in a variety of contexts and introduce the concepts of *subsample, significance level,* and *type I* and *type II* errors and the losses associated with them.

9.1 Discrimination

When the affected individuals number in the hundreds, the courts have little difficulty in discerning a pattern of discrimination.

In *Yick Wo v. Hopkins*,[3] an ordinance prohibited operation of 310 laundries that were housed in wooden buildings, but allowed such laundries to resume operations if the operators secured permits from the government. When laundry operators applied for permits to resume operation, all but one of the white applicants received permits, but none of more than 200 Chinese applicants was successful.

A state legislature violated the Fifteenth Amendment by altering the boundaries of a particular city "from a square to an uncouth twenty-eight-sided figure."[4] The alterations excluded 395 of 400 black voters without excluding a single white voter.

The court found the statistical disparities in these two cases "to warrant and require"[5] a "conclusion irresistible, tantamount for all practical purposes to a mathematical demonstration,"[6] that the state acted with a discriminatory purpose.

9.1.1 Eight Is Not Enough

Several courts have ruled, as a matter of law, that discrimination may not be proved by statistics involving too small a pool. "The problem with small labor pools is that slight changes in the data can drastically alter appearances."[7] In *Mayor of Philadelphia v. Educational Equality League*,[8] statistical evidence regarding a 13-member panel was insufficient, based on the small size of the sample, to support an inference of racial discrimination. In *Haskell v. Kaman Corp.*,[9] the Second Circuit ruled that ten terminations over an 11-year period was an insufficient sample size to support an inference of age discrimination. In *Coble v. Hot Springs School District No. 6*,[10]

---

[3] 118 U.S. 356 (1886).

[4] *Gomillion v. Lightfoot*, 364 U.S. 339 (1960).

[5] *Yick Wo v. Hopkins, supra*, at 373.

[6] *Gomillion v. Lightfoot, supra*, at 341.

[7] *Sengupta v. Morrison-Knudson Company, Inc.*, 804 F.2d 1072, 1076 (9th Cir. 1986).

[8] 415 U.S. 605, 621 (1974).

[9] 743 F.2d 113, 121 (2nd Cir.1984).

[10] 682 F.2d 721, 733-734 (8th Cir. 1982).

the Eighth Circuit stated that fifteen decisions over eight years was an insufficient sample size to support an inference of gender discrimination.

The Sixth Circuit found sample sizes of 13, 14, and 17 "too small for any probative value,"[11] "inherently suspect,"[12] and "unreliable."[13] Only 19 employees renders "statistical evidence practically meaningless."[14] In a D.C. Circuit case, even a sample of 35 employees was ruled inadequate.[15] In *Fisher v. Wayne Dalton Corp.*,[16] 40 positions were eliminated including those of five of the six oldest workers; still the Tenth Circuit found the statistical evidence "unavailing," in part because of the small sample size,[17] and granted summary judgment to the company.

9.1.2 Timely Objection

In *Chaves County Home Health Serv., Inc. v. Sullivan*, the District of Columbia court noted that although appellants "repeatedly emphasize that the sample sizes were too small, [they] failed to make any such objections to the statistical validity of the extrapolation in the proceedings below."[18]

9.1.3 Substantial Equivalence

We saw in Section 1.2 that the courts require a sample to be drawn from a population germane to the issue under adjudication. As the following case illustrates, not only must the sample be sufficiently large, but each of its members must be germane.

A plaintiff need not prove that pay disparity is motivated by an intention to discriminate on the basis of gender. A violation occurs when an employer pays lower wages to an employee of one gender than to substantially equivalent employees of the opposite gender in similar circumstances.[19] The burden is on the plaintiff to establish:

1. The referenced employees of the opposite gender are substantially equivalent.
2. Sufficiently many substantially equivalent employees are available to establish a pattern.

---

[11] *Tinker v. Sears, Roebuck & Co.*, 127 F.3d 519, 524 (6th Cir. 1997).

[12] *Brocklehurst v. PPG Industries*, 123 F.3d 890, 897 (6th Cir. 1997).

[13] *Simpson v. Midland-Ross Corp.*, 823 F.2d 937, 943 (6th Cir. 1987).

[14] *Coleman v. Prudential Relocation*, 975 F. Supp. 234, 240 (W.D. N.Y. 1997).

[15] *Denby v. Washington Hospital Center*, 431 F. Supp. 873 (D.D.C. 1977). See Section 9.3.1.

[16] 139 F.3d 1137 (7th Cir. 1998).

[17] Ibid. at 1140.

[18] 931 F.2d 914, 921 (D.C. Cir. 1991), <u>cert. denied</u>, 502 U.S. 1091 (1992).

[19] *Tomka v. Seiler Corp.*, 66 F.3d 1295, 1310 (2nd Cir. 1995).

In *Pollis v. New School for Social Research*,[20] the plaintiff based her case of gender discrimination on a statistical analysis of the eight tenured graduate faculty professors, six male and two female, who were subject to the mandatory retirement provisions of the New School's bylaws between 1974 and 1993.

The Second Circuit appeals court found that Pollis' statistical evidence suffered from several serious flaws that rendered it insufficient to sustain a reasonable inference that her treatment by the New School was motivated by discriminatory intent.

First, the size of the group subjected to statistical analysis was tiny — especially considering the comparisons encompassed a 20-year period.

> A statistical showing of discrimination rests on the inherent improbability that the institution's decisions would conform to the observed pattern unless intentional discrimination was present. The smaller the sample, the greater the likelihood that an observed pattern is attributable to other factors and accordingly the less persuasive the inference of discrimination to be drawn from it.[21]

In addition to the small size of the group to which Pollis sought comparison, each of the male members of the group differed so substantially from Pollis that the Second District appeals court felt no meaningful inference might be drawn from the statistics.

Three of the male faculty members included in the comparison group reached the mandatory retirement age between 22 and 27 years before Pollis. Their appointments to full professorships at that time were recommended by a different university president, and ratified by a board of trustees from which few or no members remained to participate in the later retirement decisions.

> Even more significant than the difference in decisionmakers, furthermore, is the huge lapse of time separating the decisions in question and hence the different circumstances in which the decisions were made. The evidence showed that during the early 1970s the New School had a far more permissive attitude generally toward extending teaching careers past age 70 than in 1993.
>
> Thus, Pollis' statistical comparison of her case in 1993 with three grants of full-time teaching positions to males in the late 1960s and early 1970s furnishes no useful information as to whether the decision in her case was motivated by bias.

---

[20] 132 F.3d 115 (2nd Cir. 1996).
[21] Ibid. at 121.

The statistical comparison involving five professors who became subject to the mandatory retirement policy in the 1980s and 1990s is no more probative of discriminatory intent. First, needless to say, a sample size of five is even smaller and therefore less probative than a sample of eight. Furthermore, the three males who were offered full-time positions in the 1980s and '90s were so different from Pollis, and from Henle, that the comparisons are virtually meaningless. Bruner and Heilbroner were internationally celebrated stars, who brought prestige to the institution and attracted students and other professors.[22]

In short, because her statistical group was so tiny, was spread over such a long period, and was composed largely of individuals who were not fairly comparable to her, Pollis' statistics did not support an inference about the school's motivations in offering her less than full-time employment when she reached the mandatory retirement age. The appeals court ruled against her.

9.1.4 Other Related Discrimination Opinions

In *American Federation of State, County, and Municipal Employees, AFL-CIO v. Washington*,[23] comparison of wages in dissimilar jobs was insufficient to establish an inference of discriminatory intent.

In *Smith et al. v. Virginia Commonwealth University*,[24] sample sizes were sufficiently large that plaintiff could use multiple regression as described in Chapter 12 to compare dissimilar populations.

9.2 The 80% Rule

Federal agency guidelines for the establishment of statistical proof in discrimination cases require a showing that the protected group is selected at less than four-fifths or 80% of the rate achieved by the highest scoring group.[25] With small samples, this showing alone is not enough to provide proof of discrimination.[26]

---

[22] 132 F.3d 115 (2nd Cir. 1996).

[23] 770 F.2d 1401, 1407 (9th Cir. 1985).

[24] *Smith et al. v. Virginia Commonwealth University*, 84 F.3d 672 (4th Cir. 1996).

[25] 28 C.F.R. 50.14 at 4(d) (1977). 29 CFR §1607.4(d)(1983) adopted in *Connecticut v. Teal*, 102 S.Ct. 2525, 2529, n. 4 (1982).

[26] See Section 10.4.1.

9.2.1 Differential Pass and Promotion Rates

Application of the 80% rule and possible counter-arguments are amply illustrated in *Bouman v. Block.*[27] Susan Bouman filed a suit alleging gender discrimination on behalf of herself and a class of potential female applicants for the position of sergeant with the Los Angeles County Sheriff's Department.[28]

> Of the 79 women who took the 1975 written test, ten, or roughly 13 percent, scored high enough on a combination of written and appraisal scores to be considered candidates for promotion. Four women, or roughly five percent of the women who took the examination, were ultimately promoted. Of the 1312 men who took the 1975 written test, 250, or approximately 19 percent, received sufficiently high combined scores to be eligible for promotion. 127 men, or approximately ten percent, were ultimately promoted. These figures clearly show a violation of the "80 percent rule."[29]

The women's pass rate — the number of persons placed on the eligibility list over the number who took the test, was 66% of the men's pass rate, while the women's promotion rate — the number of people promoted over the number who took the test, was less than 53% of the men's promotion rate.

The results of the 1977 examination were similar. Of the 102 women who took the 1977 written test, 18, or roughly 18%, scored high enough on a combination of written and appraisal scores to be considered candidates for promotion. Five women, or roughly 5% of the women who took the examination, were ultimately promoted. Of the 1259 men who took the 1975 written test, 331, or approximately 26%, received sufficiently high combined scores to be eligible for promotion. Ninety-three men, or approximately 7%, were ultimately promoted. These figures show a violation of the 80% rule for both the 1977 examination and the promotions based on it. The women's pass rate was only 67% of the men's pass rate, while the women's promotion rate was only 66% of the men's promotion rate.

Considering these figures, the district court ruled that the county had discriminated against Bouman and the class plaintiffs on the 1975 and 1977 sergeant's examinations.

---

[27] 940 F.2d 1211 (9th Cir. 1991).
[28] Violations of 42 U.S.C. 1983, Cal. Gov. Code 12900 and Title VII.
[29] Ibid. at 1225.

9.2.2 Sample versus Subsample Size

Los Angeles County based its appeal from this decision, in part, on a prior decision of the Ninth Circuit appeals court in *Contreras v. City of Los Angeles*[30] that had criticized the small size of the available sample. The statistical significance of a disparate impact showing in that case was undermined by the fact that if only three more members of the plaintiff group (Spanish-surnamed applicants) had passed the examination there would have been no violation of the 80% rule.[31]

> The County argues preliminarily that plaintiff must show uncontroverted evidence to establish disparate impact. This argument is without foundation. *Contreras v. City of Los Angeles* held only that where the evidence is uncontroverted, a *prima facie* case is established.[32] It established no requirement that statistical evidence be uncontroverted to establish a *prima facie* case. Moreover, after trial we review whether the verdict was supported by substantial evidence, not whether the plaintiff established a *prima facie* case sufficient to withstand pre-trial judgment.[33]

> We have criticized the Federal Agency guidelines,[34] noting that they were not promulgated as regulations and do not have the force of law. Rather than using the 80-percent rule as a touchstone, we look more generally to whether the statistical disparity is "substantial" or "significant" in a given case.[35] Nonetheless, while the guidelines are not necessarily dispositive, they are instructive.

> The trier of fact must consider the statistics in light of all the evidence.[36] Whether the statistics are undermined or rebutted in a specific case would normally be a question for the trier of fact.[37]

> The County argues that the violation of the 80-percent rule is not sufficient to support a finding of disparate impact under *Clady* because, according to the County, the numbers involved are too small to yield statistically significant results.[38] We agree

---

[30] 656 F.2d 1267 (9th Cir. 1981).

[31] See Section 9.3.

[32] Id. at 1275.

[33] *U.S. Postal Service Board of Governors v. Aikens*, 460 U.S. 711, 713 (1983).

[34] See *Clady v. County of Los Angeles*, 770 F.2d 1421, 1428 (9th Cir. 1985).

[35] Id. at 1428-1429, (citing *Contreras*, 656 F.2d at 1274-1275).

[36] See *Anderson v. Bessemer City*, 470 U.S. 564 at 573-574 (1985).

[37] Compare *Anderson*, 470 U.S. at 573 with *Teamsters*, 431 U.S. at 339-340.

[38] But see Section 9.4. Statistical significance and its interrelations with sample size are discussed in Chapter 10.

as a general matter that a violation of the 80-percent rule is not always statistically significant. In this case, however, the plaintiffs have demonstrated that the differences in the performances of men and women are statistically significant. Plaintiff's experts showed by several generally accepted techniques that the adverse impact of the examinations and the bottomline adverse impact were statistically significant.[39]

The County contends that the district court should not have credited the disparate impact data because a small number of women passed the tests and were promoted. The County points to our statement in *Contreras* that the statistical significance of a disparate impact showing in that case was undermined by the fact that if only three more members of the plaintiff group (Spanish-surnamed applicants) had passed the examination there would have been no violation of the 80-percent rule.[40] The County correctly points out that if only one additional woman had been promoted as a result of the 1977 examination, there would have been no violation of the 80-percent rule for that year. The same would be true for 1975 if just three more women had been promoted as a result of that year's examination.

In our view, the County misinterprets the significance of our statement in *Contreras*. In *Contreras*, not only was the number of people in the plaintiffs' group who succeeded on the examination small, the number who took it was small as well. Only 17 Spanish-surnamed applicants took the examination in question in *Contreras*,[41] whereas in the present case 79 women took the 1975 examination and 102 women took the 1977 examination. Generally, it is the combination of small sample size and small success rate that calls into question the statistical significance of a violation of the 80-percent rule. Moreover, in *Contreras*, there was no showing of statistical significance at the .05 level. Here, there was.[42] Such a showing indicates that — taking into account the effect of the small numbers — the disparity is statistically significant.

The County nonetheless criticizes the finding of statistical significance because it is based in part on combining the results

---

[39] Id. Appendix A.1.
[40] 656 F.2d at 1273 and n. 4.
[41] Id. at 1273.
[42] See ante, at n. 1. See also Chapter 10.

of the 1975 and 1977 examinations to yield the significance data. Yet, the courts have repeatedly looked at trends from past examinations to see if the total pass rate showed evidence of discrimination.[43] Moreover, the County's own experts at trial aggregated data from the two exams because, as one of them stated, it produces a "more powerful test"[44] and increases the number of observable cases. Bouman's aggregation of the 1975 and 1977 examinations was therefore permissible.[45]

The County also criticizes the disparate impact analysis appellee submitted to the trial court because women who were eligible to take the examination but did not actually take it were included in the pool for analysis. We need not decide whether such evidence should have been admitted, because even if the analysis is limited only to actual test takers, the aggregate promotion rates for 1975 and 1977 show a statistically significant violation of the 80-percent rule.[46]

9.3   No Sample Too Small

Absolutes can be dangerous; between them, the courts and statisticians have evolved several rules for determining on a case-by-case basis whether a sample is large enough. The Supreme Court remains divided on the issue.

9.3.1   Sensitivity Analysis

Some courts have adopted a sensitivity or "change one or two" rule in which it is not the size alone of the sample that is determinate, but whether a change in the status of one or two of the observations would affect the results.

For example, Charles and Linda Oliver were discharged by their employer Pacific Northwest Bell for "dishonest acts" they'd committed outside of their employment. The Olivers claimed this policy of their employer was discriminatory because of its disproportionate impact on blacks. Of 18 individuals whom Bell discharged for this reason, 6 or 33.3% were black, yet only 4.6% of Bell employees were black. The court found

---

[43] See *Ezell v. Mobile Housing Bd.*, 709 F.2d 1376, 1382 (11th Cir. 1983); *Boston Chapter NAACP v. Beecher*, 504 F.2d 1017, 1021 (1st Cir. 1974).

[44] See Section 13.3.1.

[45] 940 F.2d 1211 at 1226 (9th Cir. 1991).

[46] 28 C.F.R. 50.14 at 4(d) (1977).

the sample too small to indicate disparate impact, because the subtraction of even one or two blacks, including the Olivers, would have changed the percentage discharged significantly.[47]

Denby v. Washington Hospital Center[48] involved 9 nonwhites out of a sample of 35; the 80% rule applied; regardless, Judge Sirica noted that

> With so meager a sample, if just a handful of test results had turned out differently, the comparative percentages of black (44%) and white (100%) success on the exam would have been correspondingly, and substantially, different.

In a reduction in force case that entailed potential age discrimination,[49] if only two of the 13 employees in their thirties who were not placed at risk for discharge were switched with two employees in their forties who were placed at risk, the ages of the employees who were terminated would be almost identical to the ages of the employees who were not. The Eighth Circuit agreed with the district court that the plaintiff should not be allowed to argue before a jury that these small numbers somehow implied age discrimination.

In a similar case before the Tenth Circuit, the plaintiff had prevailed at trial, but the appeals court reversed with regard to the plaintiff's age statistics, instructing that:

> The group of non-managerial geologists over forty, which consists of only nine geologists, is too small to provide reliable statistical results. Random fluctuations regarding the retention or termination of just one or two geologists within this group during the March 1986 reduction in force would have had an enormous impact on the percentage of geologists over forty who survived the reduction in force. Consequently, such a small statistical sample carries little or no probative force to show discrimination.[50]

Similar results were reached in *Bridgeport Guardians Inc. v. Members of Bridgeport Civil Service Commission*,[51] *Mayor of Philadelphia v. Educational Equality League*,[52] and *Wade v. New York Telephone Co.*[53] See Kadane [1990] (discussed in Section 13.3) for a criticism of this approach,

---

[47] *Oliver v. Pacific Northwest Bell Telephone Co.*, 106 Wash. 2d 675, 724 P.2d 1003 (1986).

[48] 431 F. Supp. 873 (D.D.C. 1977).

[49] *Garner v. Arvin Industries, Inc.*, 885 F. Supp. 1254 (E.D. Mo. 1995), aff'd, 77 F.3d 255 (8th Cir. 1996).

[50] *Fallis v. Kerr-McGee Corp.*, 944 F.2d 743, 746 (10th Cir. 1991).

[51] 354 F. Supp. 778 (D. CN.), modified, 482 F.2d 1333 (2nd Cir. 1973).

[52] 415 U.S. 605, 621 (1974).

[53] 500 F. Supp. 1170 (S.D. N.Y. 1980).

as well as *Chicano Police Officers Association v. Stover*,[54] which we discuss along with *Denby* in the next chapter. In this latter case, a sample of 26 Chicanos out of a total of 90 applicants was ruled adequate.

9.3.2 Statistical Significance

From a statistician's point of view, no sample is too small; one can always establish a significance level, but whether it will convince the court is quite another matter. Consider the following hypothetical example. Suppose the government has mandated a median level of a certain pollutant no greater than 1.5 ppm and a maximum level no greater than 2.5 ppm. On April 1, a level of 1.6 ppm is recorded just outside the Baker Company plant. Baker responds, "So what? The 1.5ppm is a median, a 50th percentile, and thus half the time pollutant values can be expected to be greater."

On May 1, the recorded pollutant level is 1.7 ppm; again, the company says, "So what? It's just like flipping a coin and getting heads twice in a row; something like this can happen 50% × 50% = 25% of the time."

On June 1, the pollutant level is 1.8 ppm and the government notes that the odds of these three successive high readings occurring by chance alone, when the true median level is less than 1.5 ppm, are less than 50% of 25% or 12.5%. (This is also the probability of throwing heads three times in succession.)

On July 1, the pollutant level is 1.9 ppm again, and the government goes to court. It claims that an event that occurs by chance less than 50% of 12.5% or 6.25% of the time is statistically significant. (This is also the probability of throwing heads four times in succession.) Baker says, "No, a probability value (or p-value) has to be less than 5% to be statistically significant; besides, if just one of the readings had gone the other way, the results would be entirely different."

Is the result significant? We need some definitions. Before a statistician performs a statistical test, a significance level must be specified. The significance level is the probability that if nothing is wrong, an alarm may be triggered by chance. A false alarm of this sort is called a *Type I error*. In establishing a significance level, a balance must be struck among four factors:

1. The cost of making a Type I error
2. The cost of making a Type II error, that is, of concluding there is no violation when a violation does exist
3. The probability of making a Type II error
4. The cost of taking a sample

---

[54] 526 F.2d 431 (10th Cir. 1975).

In this example, the cost of making a Type I error is the sum of the costs of closing down and/or modifying the Baker plant; the cost of making a Type II error is the sum of the many costs associated with excessive pollution in the atmosphere.

Of course, these factors and the weight to be given each should have been spelled out in the anti-pollution regulation and not left to the courts to resolve. Typically, a significance level of 2% might have been specified, requiring six successive high readings before an offending plant was ordered closed. (This would be the equivalent of throwing six heads in a row with a fair coin.)

How the significance level is determined, the interrelationship between sample size and significance level, and its application by the courts form the substance of the next chapter.

9.3.3 Collateral Evidence

Small sample sizes also can be offset through the use of collateral (and non-statistical) evidence. See *Williams v. City and County of San Francisco*.[55] The First Circuit continues to affirm that statistical evidence is admissible regardless of sample size. "The reliability of his [statistical] analysis affect[s] not the admissibility of his statistical testimony, but the weight which the jury might choose to give it."[56]

> Even small samples are not per se unacceptable. See *Fudge v. Providence Fire Dep't*.[57] .... The probative worth of statistical testimony must be evaluated in the light of the methodology employed, the data available, and the factual mosaic unique to the case at hand.[58]

> While we appreciate that "small sample size may ... detract from the value of [statistical] evidence,"[59] a defendant who asserts that a plaintiff's *prima facie* case is insufficient must point out real deficiencies ...."[60,61]

---

[55] 483 F. Supp. 335, 341-342 (N.D. Cal. 1979).

[56] *Freeman v. Package Machinery Company*, 836 F.2d 1331, 1338 (1st Cir. 1988).

[57] 766 F.2d 650, 658 (1st Cir. 1985).

[58] *Freeman* at 1342, fn. 5.

[59] *Teamsters v. U.S.*, 431 U.S. 324, 339 n. 20 (1977).

[60] *E.E.O.C. v. Steamship Clerks Union, Local 1066*, 48 F.3d 594, 604 (1st Cir. 1995).

[61] We consider further rulings on this topic in Section 10.4.1.

9.3.4 Supreme Court Division

The issue of how many is enough has divided the Supreme Court. In *Rose v. Mitchell*,[62] the defendants, all African-Americans, appealed their guilty verdict on the basis that African-Americans had been prevented from serving as jury foremen. The majority opinion stated:

> There was no evidence as to the total number of foremen appointed by the judges in Tipton County during the critical period of time. Absent such evidence, it is difficult to say that the number of Negroes appointed foreman, even if zero, is statistically so significant as to make out a case of discrimination under the "rule of exclusion." The only testimony in the record concerning Negro population of the county was to the effect that it was approximately 30%. Given the fact that any foreman was not limited in the number of 2-year terms he could serve, and given the inclination on the part of the judge to reappoint, it is likely that during the period in question only a few persons in actual number served as foremen of the grand jury. If the number was small enough, the disparity between the ratio of Negroes chosen to be foreman to the total number of foremen, and the ratio of Negroes to the total population of the county, might not be 'sufficiently large [that] it is unlikely that [this disparity] is due solely to chance or accident.'[63]

> Inasmuch as there is no evidence in the record of the number of foremen appointed, it is not possible to perform the calculations and comparisons needed to permit a court to conclude that a statistical case of discrimination had been made out, and proof under the 'rule of exclusion' fails.[64]

In their dissent, Justices White and Stevens wrote:

> This case involves only the foreman, rather than the entire grand jury, does have implications for the manner in which respondents may meet their burden of proving discrimination. In the context of racial discrimination in the selection of juries, "the systematic exclusion of Negroes is itself such an unequal application of the law ... as to show intentional discrimination," a necessary component of any equal protection violation.[65]

---

[62] 443 U.S. 545 (1979).

[63] *Castaneda v. Partida*, 430 U.S. 482, 494 n. 13 (1977).

[64] Citations omitted.

[65] *Washington v. Davis*, 426 U.S. 229, 241 (1976).

Generally, in those cases in which we have found unconstitutional discrimination in jury selection, those alleging discrimination have relied upon a significant statistical discrepancy between the percentage of the underrepresented group in the population and the percentage of this group called to serve as jurors, combined with a selection procedure "that is susceptible of abuse or is not racially neutral."[66]

Once this showing is made, the burden shifts to the State to rebut the inference of discriminatory purpose. This method of proof, sometimes called the "rule of exclusion," may not be well suited when the focus of inquiry is a single officeholder whose term lasts two full years, as is true of the Tipton County grand jury foreman [in the present case]. For instance, in *Castaneda v. Partida*, we considered statistics relating to an 11-year period showing that 39% of the 870 persons selected for grand jury duty were Hispanic, from a general population that was over 79% Hispanic. The likelihood that this statistical discrepancy could be explained on the basis of chance alone was less than 1 in 10140. The sample size necessarily considered in a case of discrimination in the selection of a foreman simply does not permit a statistical inference as overwhelming as that in *Castaneda*. During any 11-year period, there would be only five or six opportunities for selecting jury foremen in Tipton County, assuming that every foreman selected serves at least the full 2-year term.[67]

The key numbers to compare are the number of blacks selected to be foremen and the total number of opportunities to select a foreman. The latter number may be greater than the number of different individuals who serve if the appointing judge has an inclination to reappoint those who have previously served.

Despite the inherent difficulty of any statistical presentation with respect to discrimination in filling a particular grand jury spot, respondents nonetheless have made a strong showing of underrepresentation supporting an inference of purposeful discrimination. This Court is not in a position to reject the finding, explicitly made by the Court of Appeals and implicitly made

---

[66] *Castaneda v. Partida, supra*, at 494. See, e.g., *Alexander v. Louisiana, supra*; *Turner v. Fouche*, 396 U.S. 346 (1970); *Carter v. Jury Comm'n*, 396 U.S. 320 (1970).
[67] Citations omitted.

by the District Court, that those who testified believed there had never been a black foreman during the period 1951–1973.[68] Assuming that 11 foreman selections were made during this period, the expected number of black foremen would be more than 3 — and the likelihood of no blacks being chosen would be less than 1 in 50 — if blacks, who constituted nearly a third of the county's population, and whites had an equal chance of being selected. I do not see how respondents could be expected to make a stronger statistical showing.

9.4 Summary

Evidence based on a sample must include information about the size of the sample and the manner in which the sample was selected.[69]

In a discrimination case, the composition of the sample should be comparable (age, race, sex, years of experience) to that of the plaintiff or class of plaintiffs in all aspects but the one at issue.

The courts have adopted several different rules for determining whether sample size is adequate:

■ The 80% rule
■ The leave-out-one-or-two rule
■ Statistical significance

These rules may be applied alone or in combination. Generally, it is the combination of small sample size and small success rate that calls into question the 80% rule.

Government regulations should be drafted so as to specify either the sample size and cut-off criteria or acceptable values for Type I and Type II errors.

---

[68] Citations omitted.
[69] *Florida Bar v. Went For It, Inc.*, 515 U.S. 618 (1995); see Chapter 3.

Chapter 10

Methods of Analysis

Courts ... from time to time have used straight percentage comparisons without the necessary standard deviation analysis in proving and rebutting discrimination cases. Statisticians do not simply look at two statistics ... and make a subjective conclusion that the statistics are significantly different. Rather, statisticians compare figures through an objective process known as hypothesis testing.[1]

In this chapter, you will learn a variety of methods for data analysis and of the courts' mixed acceptance thereof. You will learn the concept of significance level and its relation to sample size and the underlying population. You will learn both distribution-dependent and distribution-free methods for testing hypotheses.

10.1   Comparing Two Samples

In this section, we study a variety of statistical tests for comparing two samples and the courts' reaction to them. Each test relies on a certain set of assumptions; thus, any application of a test must include a demonstration of the validity of its assumptions.

In a discrimination hearing, one often wishes to test a null hypothesis that two samples are drawn from the same population (or from populations with identical characteristics) against the alternative that one of the two

---

[1] *Moultrie v. Martin*, 690 F.2d 1078, 1082 (4th Cir. 1982).

populations has been treated in a discriminatory fashion. In *Capaci v. Katz & Besthold, Inc.*,[2] the plaintiff complained that she and other women in the company had been discriminated against in terms of the time required to achieve promotion. The data in Table 10.1 was offered in evidence:

Table 10.1   Time in Months from Hire to Promotion[3]

Women: 229, 453

Men: 5, 7, 12, 14, 14, 14, 18, 21, 22, 23, 24, 25, 34, 37, 47, 49, 64, 67, 69, 125, 192, 483

Our first test is a permutation test. Our test statistic is the sum of the observations for the two women, 229 + 453 = 682.[4] There are 24 observations in the table, 22 for men, and two for women. If our null hypothesis is true and there are no real differences between the time to promotion for men and women, then the sum of any other pair chosen from the 24 observations should yield approximately the same result, like 5 + 453 = 458, or 5 + 7 = 12. Actually, out of the 24 × 23 = 552 possible pairs of observations, only three pairs, or 0.5%, yield a sum as large as the one observed originally. It seems unlikely that chance alone is responsible for the difference.

A good attorney for the employer would respond that this low percentage is simply the result of including too few women in the sample (see Chapter 9) and that by chance alone these women had to wait an exceptionally long time for promotion. A possible response on the plaintiff's part is to offer to replace the actual observations by ranks as shown in Table 10.2.

Table 10.2   Time to Promotion, Ranked

Women: 22, 23

Men: 1, 2, 3, 4, 5, 6, 7, 8, 9, 10, 11, 12, 13, 14, 15, 16, 17, 18, 19, 20, 21, 24

Our new test statistic is the sum of the ranks of the two observations associated with women or 22 + 23 = 45. Four possible pairs among the

---

[2]   525 F. Supp. 317 (E.D. La. 1981).

[3]   We have begun by ordering the data allowing us to see quickly that the time to promotion for the two women exceeds, by many months, the times required for all but one of the men. Even if you use a computer for the actual analysis, ordering the data should always be your first step because of the insight it provides. A graph is also a must in more complex cases; most good statistics packages will do both for you.

[4]   This statistic is easier to compute, yet yields precisely the same p-value as the more commonly used difference of the two sample means.

552, or 0.7%, have rank sums this large.[5] As statisticians, we would reject the null hypothesis at the 1% significance level.

10.1.1   A One-Sample Permutation Test

In *Moultrie v. Martin*,[6] the Fourth Circuit appeals court wanted to determine whether blacks had been discriminated against in jury selection. A comparison of the proportion of blacks in the relevant population and blacks serving on juries each year for seven years had been made using the statistic known as Student's t or simply as t.[7] If there is no real bias against blacks, we are as likely to observe a positive value of the t statistic (denoting a greater proportion of blacks on juries than in the population from which they were selected) as a negative value. The following seven values were observed: −3.4, −.9, −.9, .1, .1, −1.4, −1.8. If the signs (plus or minus) of these seven t values occurred at random, there are 2^7 or $2 \times 2 \times 2 \times 2 \times 2 \times 2 \times 2$ or 128 different ways in which they might have been chosen with only four of these ways yielding results in which blacks were as or more underrepresented than the values that were actually observed. These permutations are:

$$-3.4, -.9, -.9, \quad .1, \quad .1, -1.4, -1.8$$
$$-3.4, -.9, -.9, -.1, \quad .1, -1.4, -1.8$$
$$-3.4, -.9, -.9, \quad .1, -.1, -1.4, -1.8$$
$$-3.4, -.9, -.9, -.1, -.1, -1.4, -1.8$$

By chance alone, the probability of obtaining a set of t values as extreme as those observed is $^4/_{128}$, or 3%. As statisticians, we conclude that the results reveal a pattern of discrimination.[8]

10.1.2   Permutation Tests and Their Assumptions

A permutation test is always valid statistically providing (1) the observations are independent of one another, and (2) under the hypothesis of no differences between the two groups, the so-called *null hypothesis*, observations are all drawn from the same distribution.[9] Permutation tests

---

[5]   This permutation test using ranks is also known as the Wilcoxon test.

[6]   69 F.2d 1078 (4th Cir. 1982).

[7]   See Section 10.3.2 for a formal definition.

[8]   The appeals court did not attempt to combine the seven values into a single statistic as we have done here and reached a different conclusion.

[9]   Even if the observations are not independent, a permutation test is also valid if the observations are exchangeable under the null hypothesis, that is, if we can swap labels between samples without affecting the results; see Good [2000; p24].

can be used to make 3- and k-sample comparisons and to analyze data involving several factors at several levels.[10] Permutation tests can be applied to the original observations or, if there is concern about giving undue weight to extreme values, to their ranks.

10.2 The Underlying Population

Too small a sample may be completely unrepresentative of the underlying population, a fact of which the courts have taken repeated notice (see Chapter 9). Fortunately, as a sample grows larger, it will more and more closely resemble the population from which it is drawn. How rapidly this convergence takes place depends upon two factors:

1. The population parameters we wish to estimate
2. The nature of the underlying population

We may wish to estimate only a single central value such as the population's mean or median. We may want to determine the percentage of the population that has values in excess of 98.5, to estimate certain percentiles such as P_{10}, that value which is exceeded by 90% of the population; we may also want to be sure the frequency distribution of the sample viewed in its entirety differs by at most 1% or 2% from that of the underlying population.

Figure 10.1a depicts the frequency distribution of a population whose values are closely clustered together with few or no extreme values; a distribution of the heights of adult native-born male Norwegians would look much like this. The frequency distribution of even a medium-size sample taken from this population consisting of perhaps 25 or so values would, in the majority of cases, closely resemble the population itself.

Figure 10.1b depicts the familiar bell-shaped curve or Gaussian distribution characteristic of observational errors and of any variables that represent the sum of a large number of factors, each of which makes only a small contribution to the total. As in Figure 10.1a, values are closely clustered about a single central value. While extremely large or extremely small values may exist, they are relatively few.

Figure 10.1c depicts the exponential distribution commonly observed when measuring the mean time to failure of some component of a complex system. This distribution is asymmetric. While small values are the most likely, there is a much larger probability than in the preceding diagrams of observing extremely large values.

---

[10] See, for example, Chapter 8 of Good [2001].

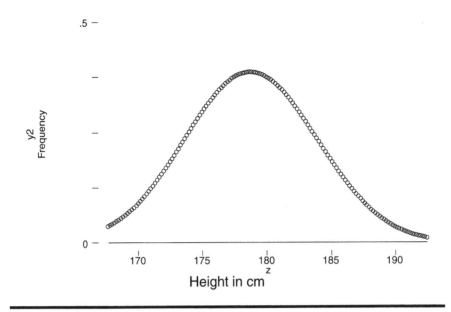

Figure 10.1a   Frequency distribution of a population with closely clustered values.

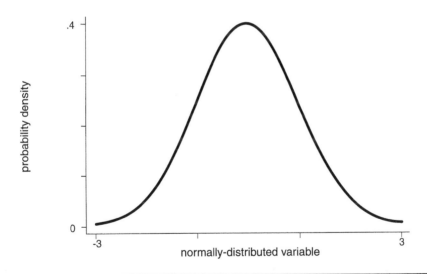

Figure 10.1b   Gaussian distribution characteristic of observational errors and variables. Values are closely clustered, with relatively few extremely large or small values.

Figure 10.1d illustrates a worst-case scenario as far as sampling is concerned. No single central value predominates and all values from the very large to the very small appear possible, if not equally likely. It would take a very large sample indeed to even begin to hint at the structure of this population.

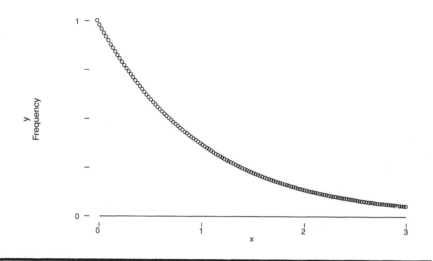

Figure 10.1c Exponential distribution commonly observed when measuring the mean time to failure (MTTF) of a component of a complex system.

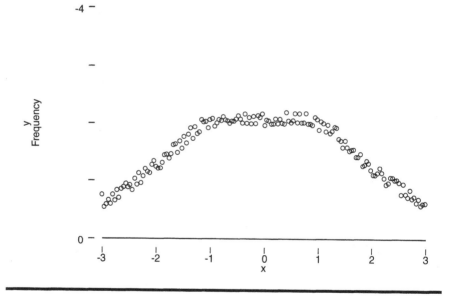

Figure 10.1d Worst-case sampling scenario — no single value predominates.

Assuming the underlying population is anything but that depicted in Figure 10.1d, we can draw a few general conclusions:

1. For any specified degree of precision, it takes a smaller sample to estimate the value of the median (or mean) than to estimate any other population parameter.

2. The smaller the variance or standard deviation of the population, the smaller the sample we need to establish statistical significance, or, for the same size sample, the smaller the difference between populations we would need.[11]

3. To determine statistical significance, differences should be expressed in standard deviations rather than absolute units. If one population is twice as variable as another, twice the difference expressed in absolute units (such as feet, dollars, or seconds) will be required to detect statistical significance and exactly the same difference expressed in standard deviations. In *Castaneda v. Partida*,[12] a case involving discrimination in jury selection, the difference between the expected number of racially balanced jury panels and the observed number was shown to be greater than two or three standard deviations.

4. It takes four times as many observations to make an estimate twice as precise. A hundred times as many observations are needed to derive an estimate ten times as precise. The precision is proportional to the square root of the sample size.

5. If we know that the variable of interest has a Gaussian, exponential, or some other well-tabulated distribution, we may be able to use tables of that distribution to determine the precision of an estimate. For example, if a distribution is Gaussian, then we know that, in drawing pairs of samples repeatedly from this distribution, a difference between the sample means as large as two standard deviations would be expected less than 5% of the time.

6. If we cannot take advantage of these tables or do not know the shape of the underlying distribution (and in most cases, we do not), we can still obtain estimates of the precision of an estimate by bootstrapping as described in Chapter 4.

A word of caution: although the appeals courts have paid some attention to these rules, as we shall see in this and succeeding chapters, district courts are more likely to fall back upon arbitrary dicta such as the 80% rule passed down from above than to make new law based upon what they view as untested statistical procedures.

10.3 Distribution Theory

In this section, we consider four theoretical distributions of which the courts have already taken judicial notice: the binomial, the normal, the Poisson,

---

[11] This point is illustrated in Figure 13.1.
[12] 430 U.S. 482 (1977).

and the exponential. We study their properties and we show how one might set up statistical tests of hypotheses concerning their parameters.

10.3.1 Binomial Distribution

The binomial frequency distribution, written B(n,p), results from a series of n independent trials each with a probability p of success, and a probability p of failure. The mathematical expectation or expected number of successes in a single trial is p. This is the proportion of successes we would observe in a very large number of repetitions of a single trial. The expected number of successes in n independent trials is pn. Thus we would expect to get five heads in 10 tosses of a fair coin.

Testing a Hypothesis

Suppose we flipped a coin in the air seven times, and six times the result was heads. Do we have reason to suspect the coin is not a fair one, that p, the probability of throwing a head, is greater than $1/2$?

To answer this question, we need to look at the frequency distribution of the binomial with n independent trials and probability p of success for each trial.

$$\Pr\{X = j\} = \binom{n}{j} p^j (1-p)^{n-j} \text{ for } j = 1, \dots$$

where $\binom{n}{j}$ denotes the number of different ways one can select j of n trials.[13]

If $n = 7$ and $j = 6$, this probability is $7p^6(1 - p)$. For $p = 1/2$, this probability is $7/128 = 0.055\%$. If six heads out of seven tries seems extreme, seven heads out of seven would seem even more extreme. Adding the probability of this more extreme event to what we have already, we see the probability of throwing six or more heads in seven tries is $8/128 = 0.0625\%$. Still, six or more heads out of seven does seem suspicious. If the next one or two throws also produce heads, it may be time to switch coins.

10.3.2 Normal Distribution

An observation that is the result of a large number of factors, each of which makes only a small contribution to the total, will have a normal

---

[13] See Chapter 4.

or Gaussian distribution. For example, if we make repeated measurements of a bookshelf's height, or the length of a flower petal, we will seldom get the same value twice if we measure to sufficient decimal places. We will get a series of measurements that are normally distributed about a single central value.

When we take the mean of a large number of bounded observations, the mean will be normally distributed or almost even if the observations come from quite different distributions.[14]

The normal distribution is symmetric; its mean, median, and mode all have the same value. To specify a normal distribution $N(\mu, \sigma)$ completely, we need only provide its mean μ and its standard deviation σ; all the rest of its properties (the shape of the frequency curve, the values of the percentiles) will be determined automatically.

In particular, fewer than 5% of the values in a normal distribution are more than two standard deviations from the mean (1.96 standard deviations to be exact). Fewer than 1% are more than three standard deviations away (3.08).

Testing a Hypothesis

A typical contract with a supplier might specify that items coming off an assembly line be 8 mm in width on average with a standard deviation of 0.024 mm. We measure the first six items coming off the line: 7.90, 8.11, 7.80, 8.01, 7.71, and 7.87. These measurements are known to be normally distributed. Their mean is 7.9. The standard deviation of the mean of six items (also known as the standard error) is $0.024/\sqrt{6} = 0.01$. Three standard deviations below the desired mean value of 8 is $8 - 0.03 = 7.97$. A mean of six items that is less than or equal to 7.97 is significant at the 1% level. Clearly, with a mean of 7.90 mm, the manufacturing process is out of control; the assembly line should be stopped and the process reset.

The Supreme Court used the criteria of two to three standard deviations in its ruling in *Castaneda v. Partida*.[15]

Student's t

In analyzing a manufacturing process, we usually have several hundred past observations to use in estimating the standard deviation of our observations. In many other practical situations, we may have little or no

---

[14] How large is large is a highly technical question beyond the scope of this book. If the observations are all of comparable size, then the mean of as few as 12 observations will be normally distributed.

[15] 430 U.S. 482, 497, n. 17 (1977).

past information to use and may be forced to estimate the standard deviation at the same time we perform our test of the mean. Thus, the resulting significance levels will be guesstimates rather than exact values.

To circumvent this difficulty, W.E. Gossett, a statistician with Guinness Brewery, proposed the following statistics:

To test whether the population mean takes the value μ_0, use the distribution of the t statistic

$$\frac{(\bar{x}-\mu_0)\sqrt{n}}{\sqrt{\sum(x-\bar{x})^2/(n-1)}}$$

To test whether the means of two normally distributed variables are the same, use the t statistic

$$\frac{(\bar{x}-\bar{y})}{\sqrt{\dfrac{\sum(x-\bar{x})^2+\sum(y-\bar{y})^2}{n_x+n_y-2}(1/n_x+1/n_y)}}$$

where $\bar{x},\bar{y}$ are the sample means, and n_x, n_y are the sizes of the first and second sample, respectively.

If the observations are (1) independent, (2) identical, and (3) normally distributed, the first of these statistics for the one-sample case has Student's t distribution with n degrees of freedom; the second, used for the two-sample comparison, has Student's t distribution with $n_x + n_y - 2$ degrees of freedom.[16] Today, of course, we no longer need to know these formulas or learn how to look up the p values in tables. A computer with most commercially available statistics packages will do the work.

For samples of eight or more observations, the t test generally yields equivalent results to the permutation tests introduced in Section 10.1, whether or not the data is normally distributed. For smaller samples, the permutation test is recommended unless you can be absolutely sure the data is drawn from a normal distribution.

10.3.3 Poisson Distribution: Events Rare in Time and Space

The decay of a radioactive element, an appointment to the U.S. Supreme Court, a case of leukemia, and an individual struck by lightning have in common the fact that they are relatively rare but inevitable events. They are

---

[16] Why Student? Why not Gossett's test? Guinness did not want other breweries to guess it was using Gossett's statistical methods to improve beer quality, so Gossett wrote under a pseudonym as Student.

inevitable, that is, if there are enough atoms, men, seconds, or years in the observation period. Their frequency of occurrence has a Poisson distribution.

The number of events in a given interval has the Poisson distribution if it is the cumulative result of a large number of opportunities, each of which has only a small chance of occurring. The interval can be in space as well as time. For example, if we seed a small number of cells into a petri dish that is divided into a large number of squares, the distribution of cells per square follows the Poisson.

If an observation X has a Poisson distribution such that we may expect an average of λ events per interval, then:

$$\Pr\{X = k\} = \lambda^k e^{-\lambda}/k! \quad \text{for } k = 0, 1, 2, \ldots$$

For example, if we expect to see a taxi go by once every 5 minutes, the probability of seeing exactly one cab in the next 5 minutes is 36.8%, but so is the probability of seeing no cabs whatever. The knowledge, one rainy day when you're on your way to the airport, that 2% of the time four or more cabs will come by in the next 5 minutes is cold comfort.

On the other hand, if we know the normal incidence of leukemia in our town is one case per year and we see four cases, we know that this too could happen 2% of the time; with 100 or more towns in our state, it probably happens each year in at least two of them.[17]

10.3.4 Exponential Distribution

Instead of counting the number of taxis or cases of leukemia in a given time interval, we can time the duration between events. This latter method has the exponential distribution $\Pr[X < t] = 1 - e^{-\lambda t}$ for $t > 0$ that is depicted in Figure 10.1c. As you can see from this figure, an exponential distribution is not symmetric; it is skewed. The mean is much larger than the median because of the presence of a small percentage of very large observations.

The half-life of an exponential distribution is constant; that is, if you expect to wait an hour for a tow truck, and one still has not arrived at the end of an hour, you can expect to wait a further hour on the average.

10.3.5 Relationships among Distributions

As we saw above, the Poisson and the exponential distributions are related. So too, are the normal and Student's t distributions. Now, imagine a system,

---

[17] We should be wary of attributing too much to such chance occurrences unless we have substantial collateral evidence. It would be like painting a bull's eye around the bullet holes. See Section 13.6.2.

one on a spacecraft for example, where various critical components have been duplicated, so that k consecutive failures are necessary before the system as a whole fails. If each component has a lifetime that follows the exponential distribution, then the lifetime of the system as a whole obeys the chi-square distribution with k degrees of freedom. The sum of the squares of k independent identically normally distributed random variables also has the chi-square distribution with k degrees of freedom.

10.3.6 Distribution-Free Statistics

Most of the distributions observed in practice are either mixtures of the basic distributions — binomial, normal, and Poisson — or can be derived from them by some kind of transformation such as that described in the preceding paragraph. The bad news is that tables such as those employed by the t statistic are not designed to be used with population mixtures, but only with known distributions. A solution is to use a distribution-free method such as a permutation test.

10.3.7 Bad Choices

Forcing a statistical test to depend upon a specific distribution can result in bad choices. In *E.E.O.C. v. Western Electric Company*,[18] the judge thought, in error, that a geometric distribution[19] should have been employed — and faulted the plaintiff — not realizing or, apparently, caring that significance would have been demonstrated with the binomial as well.

In *Branion v. Gramly*,[20] a physician convicted of murdering his wife argued that he could not have had the time to leave the hospital where he worked, drive home, garrote and shoot his wife, and report the murder to the police. In support of this contention, he offered into evidence the result of a series of computer simulations that showed the necessary driving time would be at least three standard deviations less than the average. As can be seen from Figure 10.1b, an event this many standard deviations below the average is extremely unlikely if the driving times have the Gaussian distribution. The Seventh Circuit rejected Branion's argument: "Nothing suggests a Gaussian distribution or the absence of skewness."[21]

---

[18] 713 F.2d 1011 (4th Cir. 1983).

[19] When, instead of determining the number n of binomial trials in advance, we sample until we get a predetermined number k of successes, the resulting distribution is called the negative binomial. When k = 1, as would be the case with a father anxious to have a boy after a succession of daughters, it is called the geometric distribution.

[20] 822 F.2d 1256 (7th Cir. 1988), cert. denied, 490 U.S. 1008 (1989).

[21] Id. at 1265.

If the driving times had a skewed distribution as depicted in Figure 10.1c, then a quick trip from the hospital was not at all improbable.

10.4 Contingency Tables

In almost all surveys, the results fall into categories rather than being measurable as in the preceding sections on a continuous or discrete ordinal scale. Examples of parameters include male v. female, African-American v. Hispanic v. Oriental v. Caucasian, in favor v. against v. undecided. Results can be reported in the form of a contingency table. Table 10.3, derived from *Sheehan v. Daily Racing Form, Incorporated*,[22] is an example.

Table 10.3   Retention in Employment as a Function of Age[23]

	42 years or less	48 years or more
Retained	6	2
Discharged	0	9

Following an acquisition, Jim Sheehan, the plaintiff, was discharged. He alleged that the discharge was discriminatory, based solely on age. He offered in evidence a list his former boss sent him, listing names, occupations, and birth dates of existing employees.

Of the 17 persons on the list, 11 were 48 years old or older and only 2 of these were retained. Six were 42 or younger (there were none between 42 and 48) and all were asked to remain. An affidavit by a statistician hired by the plaintiff as an expert witness stated that the probability that the retentions in the list of 17 are uncorrelated with age is less than 5%.

The statistician used Fisher's exact test to derive this probability. Let's take another look at the data of Table 10.3 to see how that was done. Table 10.4 has several fixed elements: the total number of persons 42 and under, 6, and the total number of persons 48 and over, 11, the total number retained, 8, and the total number discharged, 9.

These totals, collectively known as the marginals, are immutable; no swapping of labels will alter them. But these totals do not determine the contents of the table as can be seen from Tables 10.5 and 10.6 whose marginals are identical with those of our original table.

---

[22] 104 F.3d 940 (7th Cir. 1997).
[23] Ibid. at 944.

Table 10.4   Retention in Employment as a Function of Age

	42 years or less	48 years or more	Total
Retained	6	2	8
Discharged	0	9	9
Total	6	11	17

Table 10.5   Retention in Employment as a Function of Age

	42 years or less	48 years or more	Total
Retained	5	3	8
Discharged	1	8	9
Total	6	11	17

Table 10.6   Retention in Employment as a Function of Age

	42 years or less	48 years or more	Total
Retained	4	4	8
Discharged	2	7	9
Total	6	11	17

Note that Table 10.4 is more extreme than Tables 10.5 and 10.6 in the sense that the disparity between the retention rates of young and old individuals is greater for Table 10.4 than it is for the other two.

Fisher [1934] would argue that if the retention rates were the same for all age groups, then each of the redistributions of labels to individuals, that is, each of the N possible contingency tables with these same four fixed marginals, is equally likely, where:

$$N = \sum_{j=0}^{6} \binom{8}{j}\binom{9}{6-j} = \binom{8+9}{6} = 12{,}376.$$

$\binom{8}{j}$ is read as "8 choose j" and represents the number of different ways one can choose a set of j things from 8 different things. For example $\binom{8}{0} = 1$; $\binom{8}{1} = 8$; $\binom{8}{2} = 8 \times 7/2 = 28$. Why 28? The first person in the set could be chosen in any of eight different ways; the second person in the set could be chosen in any of seven different ways: 8 × 7 = 56. Who is chosen first and who is chosen second do not matter, so we divide by 2. In general, $\binom{8}{j} = \dfrac{8!}{j!(8-j)}$ where j! = j(j − 1)(j − 2) ... 1.

How did Fisher get the value 12,376 for N?[24] The component terms are taken from the hypergeometric distribution:

$$\sum_{x=0}^{t} \binom{m}{x}\binom{n}{t-x} \Big/ \binom{m+n}{t}$$

where n, m, t, and x occur as the indicated elements in the following 2 × 2 contingency table.

Table 10.7   2 × 2 Contingency

	Category 1	Category 2	Total
Category A	x	t − x	t
Category B	m − x	n − (t − x)	9
Total	m	n	m + n

If both age groups have the same probability of being retained, then all tables with the marginals m, n, t are equally likely, and

$$\sum_{k=0}^{m-x} \binom{m}{m-x}\binom{t}{k}$$ are as or more extreme than the original table.

In our example, $m = 6$, $n = 11$, $x = 6$, and $t = 9$, so that $\binom{6}{6}\binom{8}{6} = 28$

tables are as extreme as our original table and none is more extreme. The probability of this occurring by chance is 28/12376 < 0.0023 or less than a quarter of a percent.[25] As statisticians, we would reject the hypothesis that

---

[24] Bruce Weir reports testifying in a case in Colorado "when the defense objected that my use of Fisher's exact test was hearsay and that the prosecution needed to call Mr. Fisher to the stand." [Gastwirth, 2000, p. 88.] A citation to Fisher [1934] or to the description of this test in almost any modern statistics textbook should suffice.

[25] At this point, you're probably asking whether we expect you to do all these calculations by hand. Not at all. StatXact from Cytel Software (617/661-2011) is designed specifically for the analysis of contingency tables on an IBM-PC or PC-clone using Fisher's exact methods.

the age is not a factor in discharge, and accept the alternative that Sheehan and other older employees were discriminated against on the basis of age.

Actually, as statisticians, we first should have done what the court did and ask whether we had in hand *all* the evidence. Two other employees affected by the move were not on the list that Sheehan provided. Both were in their fifties and both were retained. The first was Shulman (Sheehan's former boss). Because Shulman supervised the office about to be closed, his head, too, was potentially on the chopping block. The second employee not on the list was Bob Sebring, director of operations, "whose value to us is obvious," Shulman had written. Thus 19, not 17, employees were involved as shown in Table 10.8.

Table 10.8   Retention in Employment as a Function of Age

	42 years or less	*48 years or more*	*Total*
Retained	6	4	10
Discharged	0	9	9
Total	6	13	19

Let us calculate Fisher's Exact test statistic for this revised table. The total number of tables with the same marginal totals is $\binom{19}{6}$ = 27,132. The number of tables that are as extreme is $\binom{10}{6}$ = 210 or 0.7% of the total, still an extremely small number. We would be justified in arguing as Sheehan did that the affidavit in combination with the fact that the list contained the ages (more precisely, the birth dates, from which ages can be readily computed) of all the employees on the list established a jury issue of age discrimination.

The Seventh Circuit concluded to the contrary, finding that although the expert had used standard statistical methods for determining whether there was a significant correlation between age and retention for the 17 persons on the list, the omission of Sebring and Shulman from the sample tested was arbitrary. The expert should have indicated the sensitivity of his analysis to these omissions.

More important, the court concluded, was the expert's failure to correct for any potential explanatory variables other than age. Completely ignored was the possibility that age was correlated with a legitimate job-related qualification, such as familiarity with computers. "Everyone knows that younger people are on average more comfortable with computers than older people are, just as older people are on average more comfortable with manual shift cars than younger people are."[26] Other problems with the expert's testimony noted by the court include:

- The expert used the wrong population for comparison, ignoring the fact that the 17 employees held a variety of jobs. Some were unionized workers with no supervisory responsibilities. Only two were supervisors — McEvoy and Sheehan. McEvoy, who was even older than Sheehan, was nevertheless retained. Of the 19 who should have been on the list, four were supervisors (Shulman, Sebring, McEvoy, and Sheehan) and the three oldest were retained.
- The expert failed to make any adjustment for variables, other than age, bearing on the decision whether to discharge or retain a person on the list.
- The expert equated a simple statistical correlation to a causal relation. His statement, "Of course, if age had no role in termination, we should expect that equal proportions of older and younger employees would be terminated," would be true only if no other factor relevant to termination is correlated with age.

These lapses, the Seventh Circuit concluded, "indicate a failure to exercise the degree of care that a statistician would use in his scientific work, outside of the context of litigation. In litigation an expert may consider (he may have a financial incentive to consider) looser standards to apply. Since the expert's statistical study would not have been admissible at trial under *Daubert*,[27] it was entitled to zero weight in considering whether to grant or deny summary judgment."[28]

10.4.1 Which Test?

In the previous chapter, we noted that the courts have adopted several different rules for determining whether sample size is adequate:

- The 80% or 4/5th rule
- The leave-out-one-or-two rule
- Statistical significance

---

[26] One could and should quarrel with this ageist comment. Modal behavior is absolutely no guide to the individual. Are we to conclude that computer science students by virtue of their ages are more knowledgeable than their professors?

[27] *Daubert v. Merrell Dow Pharmaceuticals, Inc.*, 509 U.S. 579 (1993). See Section 8.1 where this case on the admissibility of expert evidence is discussed at length.

[28] If the results are ruled not significant, the study can be challenged on the basis that it lacked the power to discriminate. See Section 13.3.1.

Another Way

Here's another way to think of this problem. If the firings were truly independent of age, i.e., selected randomly from the entire set of ages, what is the probability that all nine fired employees would come from the older group? We can test this by putting six cards marked *young* and 11 cards marked *old* in an urn, shuffling the contents, drawing out nine cards (without replacement), then recording how many of the nine were marked *old*. A computer program using the Resampling Stats language does that task many times (15,000 here) and calculates how often all nine of the fired workers were old:

URN 6#0 11#1 workers

An urn (we'll call it "workers") with 6 "0's" (young) and 11 "1's" (old)

REPEAT 15000

  Do 15,000 simulations

SHUFFLE workers workers$

  Shuffle the urn, call the shuffled urn workers$

TAKE workers$ 1,9 fired

  Take the first 9 numbers (the "fired" workers)

COUNT fired =1 old

  Count how many were old

SCORE old scrboard

  Keep score of the result

END

COUNT scrboard =9 k

  How often did we get 9 olds?

DIVIDE k 15000 prob

Result: prob = .002

Interpretation: We can see from the result that it is very unusual to select nine workers at random to fire and have all of them be old. In this computer simulation this happened only 0.2% of the time.

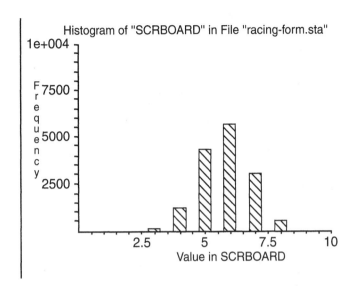

Histogram of "SCRBOARD" in File "racing-form.sta"

The last rule has been defined by the courts in a variety of ways, almost always without regard to the associated losses a statistical decision theory requires.[29] In *Segar v. Civiletti*,[30] the p value for significance was 10%. The 5% rule is endorsed by the E.E.O.C.[31] The Supreme Court set the requirement indirectly at 1 to 5% in *Castaneda v. Partida*.[32]

The 80% rule was rejected in *Fudge v. Providence Fire Dep't.*[33] in favor of tests of statistical significance.

The 80% rule described in Section 9.2 was clearly violated in *Wilmore v. City of Wilmington*; but the court ruled, "the differences in promotion rates are based on small numbers and are not statistically significant."[34]

The court was correct. In 1980, eight Caucasians out of 61 employed by the City of Wilmington were promoted, a Caucasian promotion rate of 13.1%; only one African-American out of 11 was promoted, a rate of 9.1%; whereas $4/5$ of 13.1% is 10%. Using Fisher's exact test, we can see that this event or one more extreme could happen as often as 35.7% of the time.

---

[29] If the results are ruled not significant, the study can be challenged on the basis that it lacked the power to discriminate. See Section 13.4.

[30] 508 F. Supp. 690, 700-701 (D.D.C. 1981).

[31] 29 C.F.R. 1607.14(a)(5). See also Shoben [1978; n. 41 at 803].

[32] 430 U.S. 482, 497, n. 17 (1977).

[33] 766 F.2d 650, 658, fn. 10 (1st Cir. 1985).

[34] 533 F. Supp. 844, 854 (D. Del. 1982).

In *Commonwealth v. Rizzo*, the court stated, "We will not reject the results of plaintiff's study simply by lining it up with the 5% level."[35] This court also felt the size of the sample was too small to form the basis of an opinion.

The Ninth Circuit appeals court made a ruling based on the district court's consideration of Fisher's exact test in *Kaplan v. Internat'l Alliance of Theatrical Artists*,[36] but found the evidence of specific discriminatory acts to be as or more important.[37] In *Jackson v. Nassau County Civil Service Commission*,[38] the court applied the "change one or two" rule rather than the 80% rule in concluding there was no evidence of discrimination.

In *Moore v. Southwestern Bell Telephone Co.*,[39] a test used to qualify clerks for promotion resulted in 248 of 277 African-Americans passing (a pass rate of 89.5%) and 453 of 469 Caucasians (96.6%). While this difference is highly statistically significant (<1%), the Fifth Circuit appeals court affirmed the district court's holding that a 7.1% differential between the acceptance rates of white and black applicants did not evidence the required disproportionate impact needed to make a *prima facie* case of discrimination.[40]

In *Cormier v. PPG Industries*,[41] the court considered the $4/5$ rule, significance levels based upon Fisher's exact test, correlation,[42] and the "differences in excess of two to three standard deviations" rule. "The court finds that the $4/5$ rule is not an appropriate statistical test to use in examining the question of whether blacks have been discriminated against by being denied entrance into specific job categories."

10.4.2   One Tail or Two?

When the U.S. Food and Drug Administration first imposed new regulations in the 1960s, a number of substances were given provisional approval pending further study. Among them was a food coloring, Red Dye No. 2. On looking over the results of studies in the U.S. and the Soviet Union subsequently, the Commissioner of Food and Drugs terminated provisional approval and the Certified Color Manufacturers sued.

Included in the data submitted to the court was the information on Table 10.9a. An analysis of this table by Fisher's exact test reveals a statistically significant dose response to the dye.[43]

---

[35] 466 F. Supp. 1219, 1229–1231 (E.D. PA 1979).
[36] 525 F.2d 1354 (9th Cir. 1975).
[37] Id. at 187, fn. 33.
[38] 424 F. Supp. 1162 (E.D. N.Y. 1976).
[39] 593 F.2d 607, 608, n. 1 (5th Cir. 1979).
[40] See also Section 4.3 on the folly of comparing differences in rates.
[41] 519 F. Supp. 211 (W.D. La. 1981).
[42] See Section 11.1.
[43] *Certified Color Manufacturers Association v. Mathews*, 543 F.2d 284 n. 31 (D.C. 1976).

Table 10.9a Rats Fed Red Dye No. 2

	Low dose	High Dose
No cancer	14	14
Cancer	0	7

The response is significant, that is, if the court tests the null hypothesis that Red Dye No. 2 does not affect cancer incidence against the alternative that high doses do induce cancer, at least in rats. The null hypothesis is rejected because only a small fraction of the tables with the marginals shown in Table 10.9a reveal a toxic effect as extreme as the one actually observed. This is an example of a one-tailed test. Or is it?

What would your reaction have been if the results had taken the form shown in Table 10.9b?

Table 10.9b Rats Fed Red Dye No. 2

	Low dose	High Dose
No cancer	7	21
Cancer	7	0

You would have leaped with joy, knowing that Red Dye No. 2 prevented tumors, at least in rats. Should the court have guarded against this eventuality, that is, should it have required a two-tailed test that would have rejected the null hypothesis if either extreme were observed? Probably not, but a federal district court was misled into making just such a decision in *Commonwealth of Pennsylvania et al. v. Rizzo et al.*[44]

African-American applicants for promotion in the Philadelphia Fire Department sued the city, alleging that the test was inherently unfair. The results of that test are summarized in Table 10.10.

Table 10.10 Scores on Department Examinations[45]

	Caucasians		African-Americans		
	#	Range	#	Range	Cutoff
Assistant Fire Chief	25	73–107	2	71–99	100
Fire Deputy Chief	45	76–106	1	97	100
Fire Battalion Chief	99	58–107	6	83–93	94

[44] 466 F. Supp. 1219 (E.D. Pa. 1979).
[45] Ibid. Data abstracted from Appendix A.

These results look suspicious, given that the cutoff point always seems to be just above the African-American candidates' highest scores. However, a lower cutoff point would also have included many Caucasian candidates and there were only a limited number of vacancies.

Fisher's exact test applied to the pass/fail results was only marginally significant at 0.0513; still, the court ruled, "We will not reject the result of plaintiffs' study simply by mechanically lining it up with the 5% level."[46]

Plaintiffs used a one-tailed test: does a smaller proportion of African-Americans score at or above the cutoff? The defendants used a two-tailed test: are there differences in the proportions of African-American and Caucasian candidates scoring at or above the cutoff point? The court agreed, in error, we feel, given the history of discrimination against African-Americans alone, to consider the two-tailed test as well as the one-tailed one.[47] More telling was that the mean scores of African-Americans and Caucasians on the test were not significantly different. In the end, the court applied the leave-one-or-two-out rule in making its decision.[48]

10.4.3 The Chi-Square Statistic

In this section we briefly describe the chi-square statistic although it is now considered inferior to (and should be replaced by) Fisher's exact test.[49] Moreover, the real lesson to be learned from *Smith v. Salt River*[50] is not to use the chi-square statistic or any other single-factor comparison when multiple factors can provide a more convincing explanation.

Smith alleged that because proportionately fewer African-Americans residing in the Salt River District live in owner-occupied homes, compard with non-Hispanic Caucasians, the district's land ownership voting prerequisite denied African-Americans the opportunity to participate in the district's political processes and elect representatives of their choice.

The parties entered into a lengthy joint stipulation of facts, reserving as a question for trial only the extent, if any, and statistical significance, if any, of African-Americans within the district having a lower incidence of home ownership than non-Hispanic whites.

The bench trial that ensued consisted primarily of expert testimony by Smith's statistical expert and the district's statistical expert regarding

---

[46] Id. at 1228–1229.

[47] Logically, only one of the tests should have been admitted into evidence as a test can be one-tailed or two-tailed but not both.

[48] See Section 9.3.1.

[49] See, for example, Good [2001, Chapter 7].

[50] 109 F.3d 586 (9th Cir. 1997).

the relationship, or lack thereof, between race and home ownership in the district.

Smith's expert used the chi-square approximation to analyze the demographic data. He contrasted the 60:40 ratio of Caucasian home ownership against the 40:60 ratio of African-American home ownership. The result, he testified, was "extraordinary ... the chances of finding a relationship like this between white ownership and black rentership were less than one in a million."[51]

The chi-square approximation makes use of the statistic $[O - E]^2/E$ where O is the actual number of observations in one of the cells and E is the number to be expected by chance. As the published decision includes only percentages and does not include the sample sizes, we can neither verify nor disprove the expert's conclusion. Let's see why. If there were 10 Caucasian homes and 10 African-American homes so that the four-cell or 2×2 table appeared as in Table 10.11, the statistic would be $1/5 = 0.2$. If there were 100 homes belonging to the members of each race, this statistic would take the value 2, ten times as large. The first number does not represent a statistically significant difference. The second does.

Table 10.11 Home Ownership by Race

	Caucasian	African-American
Owns own home	6	4
Rents	4	6

Preventive Statistics

Insist that your expert witnesses maintain an impartial, objective attitude. Smith's expert testified the chi-square test merely confirmed a result he expected, or as the Ninth Circuit appeals court noted, he "knew at the outset, without even calculating anything,"[52] that he would find a statistically significant relationship.

> On cross-examination, he conceded the chi-square method does not reveal how two variables will vary in relation to one another. He also conceded that he had not undertaken to identify and examine other variables that might contribute to home ownership in the District.[53]

---

[51] Id. at 590, n. 2.

[52] Id. at 590.

[53] Id.

The district's expert criticized the chi-square method as simplistic and misleading. He testified he had analyzed the same data using a multivariate model.[54] This model focuses on the variable of interest, here race, but also includes "theoretically reasonable and cogent explanations for home ownership that might compete with race."[55]

The district's expert explained in detail the operation of his model. He testified that multiple regression analysis did not indicate a strong correlation between race and home ownership. On cross-examination, he explained that his model treated race as one possible predictor of home ownership and tested whether home ownership was a function of race alone or of many factors, including race. He also stated that if forced to identify the variable with the largest net effect on home ownership, he would point to "persons per dwelling unit."

His model included ten variables: percent black, percent non-Hispanic blacks, percent of homes built since 1980, persons per dwelling unit, percent income over $50,000, percent retired or over age 65, percent at or below poverty line, percent living in same dwelling as five years ago, median household income, and percent owner occupancy. The first two variables represented race, the tenth was the dependent variable.[56]

The district court concluded that "the observed difference in rates of home ownership between non-Hispanic whites and African-Americans is not substantially explained by race but is better explained by other factors independent of race." The Ninth Circuit appeals court affirmed this decision.

10.5 Summary

Errors represent an inevitable part of any statistical procedure. A major objective is to control the probability of making a Type I error, that is, of rejecting the hypothesis when the hypothesis is true. This probability is known as the significance level of the testing procedure.

With distribution-free procedures such as the two-sample permutation test, the Wilcoxon rank test, and Fisher's exact test, the set of values of the test statistic for which we accept the hypothesis will depend only upon the significance level and the sample size. With distribution-dependent statistics such as Student's t, the set of values forming the acceptance region will also depend upon the nature of the underlying population.

---

[54] See Chapter 11.
[55] Id.
[56] Id. at 590, n. 2.

The significance level is accurate (exact) only if the assumptions underlying the test are satisfied. With almost all procedures, the observations must be independent of one another.[57]

10.6 To Learn More

For a good introduction to distribution-free methods of statistical analysis, see Good [2000, 2001]. For a glimpse at some of the problems that beset distribution-dependent methods, see Kaye [1986].

---

[57] See Section 5.4 for a formal definition of independence. Sometimes, we can transform the data to obtain independent observations. For example, today's Dow-Jones average clearly depends upon its value on the preceding business day, but the day-to-day gains and losses obtained by subtracting the successive averages may well be independent of one another.

Chapter 11

Correlation

There are three kinds of lies: lies, damned lies, and statistics.[1]

Though perhaps hyperbolic, this declaration of distrust aptly warns that any conclusion based on statistics may be unsound. It is most unfortunate, therefore, that the evidence in this case is almost entirely statistical.[2]

Statistics should be used properly and should not be used to make inferences they do not support.[3]

In the previous chapter, we saw that the courts distrust arguments predicated on the presence or absence of a single factor. In this chapter, we consider regression methods for assessing the relative contributions made by multiple factors to earnings, housing, and employment. We also consider the many counter arguments that can be and are raised.

11.1   Correlation

In *Craig v. Los Angeles County*,[4] a group of Mexican-Americans sued the county of Los Angeles claiming they were discriminated against by the

---

[1] Generally attributed to Mark Twain.
[2] *Sobel v. Yeshiva University*, 566 F. Supp. 1166 (S.D. N.Y. 1983).
[3] *U.S. v. Ironworkers Local 86*, 443 F.2d 544, 551 (9th Cir. 1971).
[4] 626 F.2d 659 (9th Cir. 1980).

sheriff department's use of an admissions examination that was irrelevant to subsequent academic performance. The district and appeals courts found to the contrary that "the ET4-1 validation study offered in this case produced a significant correlation coefficient (.60) between the test and academic performance as measured by scores on written examinations given at the end of training."[5]

The correlation R between two variables X and Y, for example, the score on an examination and subsequent academic performance as in the instant case, is defined as:

$$R = \frac{\text{Covariance}[xy]}{\sigma_x \sigma_y} = \frac{\sum (x_i - \bar{x})(y_i - \bar{y})}{\sqrt{\sum (x_i - \bar{x})^2 \sum (y_i - \bar{y})^2}}$$

where x_i, y_i denote the scores for the ith applicant, and $\bar{x}, \bar{y}$ are the mean scores of all applicants.[6] The denominator of R is the product of the standard deviations of the two variables; it scales the correlation so that it is units-free and takes values between 0 and 1.

A perfect positive correlation R would mean that every applicant for deputy sheriff who scored high on the ET4-1 examination also performed well in the academic courses at the sheriff's academy and that every candidate who did poorly in the preemployment test later did poorly at the academy.[7]

11.1.1 Statistical Significance

Statistical significance is determined by testing whether the correlation coefficient R, or ρ (rho), is significantly different from zero. Because the sample means and standard deviations do not change when the observations are permuted and the null hypothesis is true, a permutation test based only on $R' = \sum X_i Y_i$ produces results identical to a test based on the more complicated expression R.[8]

Let's see how this would work in a specific case. Suppose the observed values of X were 0, 1, 2 and the values of Y were 0, 1, and 2 also. If the observed pairs were (0,0), (1,1), (2,2), then $R' = 5$. If the observed pairs were (0,1), (1,2), and (2,0), then $R' = 2$. In the first instance, the values

[5] Id. at 664.
[6] While the courts may refer to R, most statisticians denote the correlation by the Greek letter ρ (rho).
[7] Id. n. 5; see also *Chemical Mfrs. Ass'n. v. EPA*, 870 F.2d. 177, 215, n.139 (5th Cir. 1989).
[8] As shown, for example, in Good [2000], p. 42.

of X and Y increase together; in the second instance, X and Y behave independently. A positive dependence between X and Y leads to a larger value of R' (or, equivalently, of R).

11.1.2 Practical Significance

R^2 is of greater practical significance than R as it measures the amount of variability in the second variable (in *Craig*, academic performance) that is explained by the first variable (in *Craig*, the score on the ET4-1 exam). Here is how one jurist described R^2 in reviewing an appeal of a death sentence in *McCleskey v. Kemp*:[9]

> Statisticians use a measure called an R^2 to measure what portion of the variance in the dependent variable (death sentencing rate, in this case) is accounted for by the independent variables of the model. A perfectly predictive model would have an R^2 value of 1.0. A model with no predictive power would have an R^2 value of 0. The R^2 value of Baldus' most complex model, the 230-variable model, was between .46 and .48. Thus, as the [district] court explained, "the 230-variable model does not predict the outcome in half of the cases."[10,11]

Similarly, in *U.S. v. City of Chicago*,[12] the court found that police sergeants' performance on a written promotion exam had a statistically significant correlation with their efficiency ratings on the job, but the correlation of 0.247 was not of practical significance.

11.1.3 Absence of Correlation

Absence of correlation can be telling. If a union operated in a nondiscriminatory fashion, members laid off the job first would also be the first to be referred to other jobs. In *Commonwealth of Pennsylvania v. Local Union 542, International Union of Operating Engineers*,[13] two lists were

---

[9] 481 U.S. 279 (1987).

[10] Id. at 289.

[11] This statement is misleading. While a large value of R signifies that a model provides a good fit to existing data, it provides no guarantee of future predictive power. Fortunately, in *McCleskey*, the fit or lack thereof of the model, and not its predictive power, was of primary interest.

[12] 385 F. Supp. 543 (N.D. Ill. 1974).

[13] 469 F. Supp. 329, 357 (E.D. Pa. 1978), aff'd 648 F.2d 9222 (3rd Cir. 1981), rev'd sub nom *General Building Contractors Ass'n., Inc. v. Pennsylvania*, 102 S. Ct. 3141 (1982).

compared. On the first list, members of the union were ranked by how often they were out of work. On the second, they were ranked in the order in which they were referred to jobs. Seventeen pairs of lists corresponding to different job categories and time periods were compared. If the union referred workers to jobs in the order they were laid off, each pair of lists would have been highly correlated. Instead, the values of R ranged from a high of 0.62 to a low of 0.08, which, using the formula $1 - R^2$, left as much as 61.6% to 99.4% of the total variation unexplained and thus attributable to discrimination. The court ruled the union acted in a discriminatory fashion.

11.1.4 Which Variables?

A positive correlation R for two variables X and Y does not mean a cause and effect relationship exists between the two. X and Y also will be correlated when a third variable Z affects both of them.

In *Barnes v. Glen Theatre, Inc.*,[14] the police attempted to close a strip joint on the grounds it served as a magnet for prostitutes and sexual deviants. As Justice Souter noted in his concurring opinion:

> To say that pernicious secondary effects are associated with nude dancing establishments is not necessarily to say that such effects result from the persuasive effect of the expression inherent in nude dancing. It is to say, rather, only that the effects are correlated with the existence of establishments offering such dancing, without deciding what the precise causes of the correlation actually are. It is possible, for example, that the higher incidence of prostitution and sexual assault in the vicinity of adult entertainment locations results from the concentration of crowds of men predisposed to such activities, or from the simple viewing of nude bodies, regardless of whether those bodies are engaged in expression or not. In neither case would the chain of causation run through the persuasive effect of the expressive component of nude dancing.
>
> Because the State's interest in banning nude dancing results from a simple correlation of such dancing with other evils, rather than from a relationship between the other evils and the expressive component of the dancing, the interest is unrelated to the suppression of free expression.[15]

---

[14] 501 U.S. 560 (1991).
[15] Id. at 586.

11.1.5 Consistency over the Range

If a correlation exists over only a part of a variable's range, the courts may choose to disregard it. Here is an example. In the absence of state support, differences from school district to school district in the amounts raised by local property taxes can create large disparities in the amounts spent on public schools. The appellees in *San Antonio Independent School District v. Rodriguez*[16] submitted the data shown in Table 11.1 and argued that this demonstrated a direct correlation between the wealth of families residing in a district and the amounts spent for education.

Table 11.1   Texas School Finance Data

Market Value of Taxable Property per Pupil	Median Family Income from 1960	Percent Minority Pupils	State and Local Revenues per Pupil	Number of Districts
Above $100,000	$5,900	8%	$815	10
$100,000–$50,000	$4,425	32%	$544	26
$50,000–$30,000	$4,900	23%	$483	30
$30,000–$10,000	$5,050	31%	$462	40
Below $10,000	$3,325	79%	$305	4

The district court agreed, but Justice Powell, writing for the majority, felt the evidence did not support such a conclusion:

> Professor Berke's affidavit is based on a survey of approximately 10% of the school districts in Texas. His findings, previously set out in the margin [Table 11.1], show only that the wealthiest few districts in the sample have the highest median family incomes and spend the most on education, and that the several poorest districts have the lowest family incomes and devote the least amount of money to education. For the remainder of the districts — 96 districts composing almost 90% of the sample — the correlation is inverted, i.e., the districts that spend next to the most money on education are populated by families having next to the lowest median family incomes while the districts spending the least have the highest median family incomes. It is evident that, even if the conceptual questions were answered favorably to appellees, no factual basis exists upon which to found a claim of comparative wealth discrimination.

[16] 411 U.S. 1, 46, n. 101.

11.1.6 Bias

All the defenses raised in Chapter 8 are applicable here as well. Correlation studies can be attacked on the basis of confounding, selection, response and observer bias, and changes in classification methodology. Bias due to self-selection was recognized by the court in *Washington v. Davis*,[17] where the pass rate for black candidates fell precisely because blacks had been urged to take a qualification test even if they doubted they would succeed.

11.2 Testing

Tests must be job-related or a business necessity.[18] "The best method of establishing job-relatedness is to show that the test has 'predictive validity.'"[19] This point was illustrated in Section 11.1. Predictive validity should be distinguished from construct validity and content validity.

> "Construct validity" entails prior identification of the characteristics believed successful to job performance, "content validity" if the content of the test closely duplicates the actual duties to be performed by the applicant.[20]

Entrance and promotion examinations also should be distinguished.

11.2.1 Predictive Validity

A group of female applicants who wanted to be members of the Los Angeles police force felt the department's minimum height requirement was unrelated to job performance and thus discriminated against them unfairly.[21]

> Appellants did not confine their attack to methodological disputes; they also introduced evidence of contrary findings of other studies. They discovered that the City had been involved in four other studies that investigated the relationship of height to job performance by police. Each of these studies had concluded that height was not significantly related to police job performance. One of the studies, the "Drawn Weapon Frequency

---

[17] 426 U.S. 229 (1976).

[18] *Griggs v. Duke Power Co.*, 401 U.S. 424 (1971).

[19] *Bridgeport Guardians v. Members of Bridgeport Civil Service Comm.*, 482 F.2d 1333, 1337 (2nd Cir. 1973), cert. denied, 421 U.S. 991.

[20] Id.

[21] *Blake v. City of Los Angeles*, 595 F.2d 1367, 1378 (9th Cir. 1979).

Survey," had found no correlation between police officer height and the frequency of drawing a weapon. This finding directly contradicted appellees' contention that shorter officers are more likely to use strong force, particularly force stronger than the bar-arm control hold. They also produced numerous depositions from police officials in other major American cities that have reduced or eliminated minimum height requirements for police. This testimony indicated that persons under 5'6" in height can, and are, safely and efficiently performing all aspects of police work.[22]

Appellees sought to demonstrate the job relatedness of the LAPD's physical abilities test through the results of two validation studies. The first study attempted to correlate performance on the five events used in the physical abilities test with 11 measures of success during Police Academy training.[23] The study found that four of the five events used in the physical abilities test had some significant correlation with at least seven of the 11 measures of training success. The second study concluded that performance on the physical abilities test had some significant correlation with performance of foot pursuit, field shooting, and emergency rescue simulations. Appellees also submitted affidavits detailing the procedures that were used to develop the physical abilities test.

Appellants attacked both the methodology and conclusions of the validation studies, contending that because the studies excluded persons who had failed the physical abilities test, they

---

[22] Ibid. at 1381.
[23] The five events on the physical abilities test were: wall scale (running a total of 50 yards and scaling a smooth wall six feet high); hang (running a total of 50 yards and hanging from a chinning bar, using an overhand grip, for one minute); weight drag (running 50 feet and dragging a dead weight of 140 pounds for 50 feet); tremor (running 50 yards and holding a stylus steady for 17 seconds); endurance (running as many laps around a one-eighth mile track as possible in 12 minutes). The 11 measures of success in police academy training were: academy average (combination of academic performance, performance in physical training, marksmanship, and peer evaluations); target shooting III (score received on final test of target shooting performance given in the academy); combat shooting II and III (scores received on last two tests of combat shooting performance); physical training I, II, and III (scores received in the evaluations of performance in the academy physical training exercise program at three different points during training); self-defense (evaluation of self-defense skills at academy); peer evaluations (at 8 weeks and 16 weeks); graduation (whether the officer completed the police academy training course).

were of little value in demonstrating that the physical test excluded persons who would be unlikely to succeed as police officers. Moreover, appellants noted that the LAPD had not used a preemployment physical test during the five years prior to the time when women were first permitted to apply for police officer positions. They questioned whether the physical test had been developed to test a representative sample of major or critical work behaviors as revealed by a careful job analysis. Extensive deposition testimony was presented indicating that police departments in other major American cities that do not use preemployment physical abilities tests have experienced satisfactory job performance by police. Finally, they also questioned appellees' failure to differentially validate the physical abilities test by sex.

Viewing the record in the light most favorable to appellants, we cannot conclude that appellees met their burden of justifying use of the physical abilities test as a business necessity. The fact that the LAPD hired thousands of male police officers between 1968 and 1973 without using any preemployment physical test suggests that the practice is not essential to safe and efficient job performance.[24] Moreover, the modest correlations between scores of successful candidates on the physical test and scores during academy training on peer evaluations, tests of physical ability and shooting skills, hardly establish that the physical test is so intimately related to job performance as to be a business necessity.

Although appellees' claim that 4 of the 5 events used in the physical abilities test had some significant correlation with at least 7 of the 11 measures of training success sounds impressive, a closer analysis of the study reveals far less impressive results. The 11 measures of academy performance were based only on measures of shooting ability, physical ability, self-defense skills, and peer evaluations. The magnitude of the correlations for all but the 3 measures of physical training were always less than $r = .3$, indicating that less than 9 percent of the variance in the academy performance measures was explained by performance on the components of the physical abilities test.[25]

---

[24] Footnote omitted.
[25] Id. at fn. 16.

Appellees had to demonstrate that their measures of training success are themselves significantly related to important aspects of job performance and they utterly failed to do so.[26] Nothing this court said in *U.S. v. Ironworkers Local 86*[27] supports appellees' notion that plaintiffs are required to go beyond a showing of disproportionate impact to establish a *prima facie* violation of Title VII. Our statement that the use of statistics "is conditioned by the existence of proper supportive facts" meant only that statistics should be used properly and should not be used to make inferences they do not support. Appellants here properly used statistics to establish the disproportionate impact of appellees' selection devices. They did not need to make any greater showing to establish a *prima facie* case of sex discrimination. As we stated in *Ironworkers*: "It is our belief that the often-cited aphorism, 'statistics often tell much and Courts listen,' has particular application in Title VII cases."[28]

11.2.2 Validating the Test

A group of Mexican-Americans sued Los Angeles County claiming they were discriminated against by the county on at least three separate grounds:[29]

1. An entry-level test that 33% of Mexican-Americans and only 13% of whites failed.
2. A second entry-level test that was only weakly correlated (R = 0.6) with performance at the sheriff's academy
3. Height restrictions that were manifestly unrelated to employment

Once the plaintiff has established a *prima facie* case of discrimination, the burden shifts to the employer to justify the challenged selection device as a business necessity by showing that it is significantly job-related. This process is generally referred to as "validation."[30]

[26] See Note, "Height Standards in Police Employment and the Question of Sex Discrimination," 47 S. Cal. L. Rev. 585, 598-599 (1974).
[27] 443 F.2d 544 (9th Cir. 1971).
[28] 443 F.2d at 551; see also Shoben [1978].
[29] *Craig v. Los Angeles County*, 626 F.2d 659 (9th Cir. 1980).
[30] *Dothard v. Rawlinson*, 433 U.S. 321 (1977); *Blake v. City of Los Angeles*, 595 F.2d 1367, 1378 (9th Cir. 1979).

The Supreme Court elaborated on the legal standard governing this inquiry as follows:

[D]iscriminatory tests are impermissible unless shown, by professionally acceptable methods, to be "predictive of or significantly correlated with important elements of work behavior which comprise or are relevant to the job or jobs for which candidates are being evaluated."[31]

The validation process thus involves three distinct steps. The employer must first specify the particular trait or characteristic which the selection device is being used to identify or measure. The employer must then determine that that particular trait or characteristic is an important element of work behavior. Finally, the employer must demonstrate by "professionally acceptable methods" that the selection device is "predictive of or significantly correlated" with the element of work behavior identified in the second step.[32]

One way the defendant might satisfy the requirement would be to show that the selection test correlates with measures of success in training and that those measures in turn are significantly related to job performance.

In the instant case, the sheriff's department has shown only that the ET4-1 test correlates with performance on the academic portion of academy training. Specifically, the ET4-1 test measures the applicant's reading, writing and reasoning skills. These skills are considered essential to acceptable performance in such courses as constitutional law, criminal law and procedure, and juvenile law and procedure, which are taught at the sheriff's academy.

The ET4-1 validation study offered in this case produced a significant correlation coefficient (0.60) between the test and academic performance as measured by scores on written examinations given at the end of training. However, plaintiffs enumerate several defects in the sheriff's study which they argue are enough to render the established correlation meaningless.

Plaintiffs first attack the sample utilized in the validation study as unrepresentative of the applicant pool because only persons

---

[31] 29 CFR §1607.4(c); *Albemarle Paper Co. v. Moody*, 422 U.S. 405, 431 (1975).
[32] 626 F.2d 662.

who received passing scores on the ET4-1 were included in the sample. Plaintiffs note that, as a result, no one really knows whether some who failed the test would nonetheless have successfully completed the training program. In support of this criticism of the validation study, plaintiffs cite EEOC guidelines which state that "(w)here a validity study is conducted in which tests are administered to applicants, with criterion data collected later, the sample of subjects must be representative of the normal or typical candidate group for the job or jobs in question."[33] Plaintiffs also cite the following language from *Blake v. City of Los Angeles* as supportive of their challenge:

Appellant's attacks on the methodology of the [validation] studies are more than mere quibbles. Because neither of the studies included persons shorter than 5'6", the studies are of little value in determining whether the height requirement excludes individuals who would be more likely to resort to strong force (the asserted purpose of the height requirement).

We agree with the *Blake* construction of the cited EEOC guidelines.[34] And, as we earlier noted, noncompliance with the EEOC guidelines diminishes the probative value of the defendants' validation study. But it is not necessarily fatal.

The sheriff's department demonstrated a very high correlation between scores on the ET4-1 test and academic averages at the academy. Plaintiffs point to no data which are inconsistent with defendants' validation study. Nor do they cite serious flaws in the study's data gathering or handling which might warrant the suspicion that the obtained correlation is spurious or erroneous.

Under the circumstances here, however, the demonstrated correlation is sufficiently strong to support the inference drawn by the district court after trial that persons excluded from the academy, and thus from the sample, because they failed the ET4-1 entrance exam would not have succeeded in the academic training program. We perceive no reason at this point to require the sheriff's department to hire and train a sample of failing applicants, with the attendant expense to the county and potential unfairness to the candidates so hired, in order to provide statisticians with more certain results.[35]

---

[33] 29 C.F.R. §1607.5(b)(1) (1975).
[34] See also *U.S. v. Georgia Pacific Co.*, 474 F.2d 906, 915-916 (5th Cir. 1973).
[35] 626 F.2d 665.

11.3 Linear Regression

The NAACP sued the city of Niagara Falls,[36] contending that the existing at-large system of voting operated to "dilute" black voting strength in violation of Section 2 of the Voting Rights Act, as amended.[37] They sought to replace this system with a single-member-district method of electing members to the city council, including a requirement that one district have a majority black population. A *regression analysis* was offered in evidence to demonstrate that the Caucasian majority in Niagara Falls voted sufficiently as a bloc usually to defeat African-American voters' candidates of choice.

> The district court explained that simple regression analysis measures a candidate's share of the votes received in a particular election district as a percentage of the number of voters at the polls in that district. That percentage is correlated with the racial composition of the district, measured in terms of percentage of the voting age population in that district .... A "correlation coefficient" is generated, demonstrating how consistently voter support for a candidate or group of candidates varies with the racial composition of the election districts.[38]

11.3.1 Linear Regression Defined

The linear regression of a dependent variable Y (such as earnings) on an independent variable X (such as years in business) is usually written as:

$$Y = b_0 + b_1 \, f[X] + \varepsilon$$

where the coefficients b_0 and b_1 are numeric constants, $f[X]$ is some nondecreasing function of X (examples include X^2, $\log[X]$, $\exp[X]$, and X itself), and ε (epsilon) is a random variable consisting of observation error plus dozens of other factors with zero means that are random in nature.

---

[36] *NAACP v. City of Niagara Falls*, 65 F.3d 1002 (2nd Cir. 1994).

[37] 42 U.S.C. §1973 (1988).

[38] Id. Footnote 2, Decision and Order at 13-14; see also *Jenkins v. Red Clay Consol. Sch. Dist. Bd. of Educ.*, 4 F.3d 1103, 1119, n. 10 (3d Cir. 1993), cert. denied, 512 U.S. 1252 (1994). The Second District appeals court ruled that the regression analysis did support the plaintiff's claim of bloc voting but turned down their appeal on other grounds.

In *NAACP v. City Of Niagara Falls*, X denotes the proportion of Caucasian voters in a precinct, and Y denotes the proportion of votes received by a Caucasian candidate.

B.F. Goodrich Co. v. Department of Transportation[39] provides us with a second practical example in which tire tread depth (the dependent variable) is regressed on miles driven (the independent variable).

In a hypothetical example, a wholesaler felt it lost profits because its chief supplier conspired with other distributors to fix the retail prices of its insecticides. The wholesaler showed that during the preconspiracy period its earnings could be approximated by a straight line, that is:

$$\text{Earnings} = \$27,520 + \$12.5 * \text{Year}$$

For example, in 1977, its earnings were approximately $27,520 + $12.5 * 1977 = $52,232. Extrapolating this formula forward into the period of the alleged conspiracy gave the wholesaler a basis for determining lost profits.

Estimating the Regression Coefficients

A number of procedures exist for estimating the regression coefficients b_0 and b_1. All try to minimize the residuals that remain after the values of Y predicted by the model are subtracted from the values that were actually observed. Ordinary least squares (OLS), the most common procedure, uses those values that minimize the sum:

$$\sum (y_i - b_0 - b_1 x_i)^2$$

Least squares formulas for calculating the regression coefficients are provided in *B.F. Goodrich Co. v. Department of Transportation*.[40] In our own work, we rely on an off-the-shelf statistics program that allows a computer to make the calculations for us. This enables us to focus our attention instead on whether the assumptions of this regression method are satisfied.

The OLS estimate of the regression coefficient is $b_1 = R\nabla y/\nabla x$. Thus the slope of the regression line, b_1, is significantly different from zero only if the correlation R is significantly different from zero. If $R = b_1 = 0$, then X would be of no help in predicting Y.

---

[39] 541 F.2d 1178 (6th Cir. 1976).
[40] Id. at 1201, Appendix C.

Limitations of the Regression Method

1. The underlying "cause" of Y may not be X. It may be a third variable Z that influences both Y and X.
2. Y may have multiple causes and X may represent only a small part of the equation (see Chapter 12).
3. Every relationship has both linear and nonlinear portions; the nonlinear portion becomes evident for both extremely large and extremely small values. While a regression equation may be used for interpolation within the range of known values, we are on shaky ground if we try to extrapolate, to make predictions for conditions not previously investigated.[41]
4. Relationships are not unique. If a relationship exists between two variables X and Y, then a relationship also exists between Y and any monotone function of X. If we can fit the curve represented by Model I: $Y = A + BX + \varepsilon$, we also can fit the relationships of Model II: $Y = A + B\log[X] + \varepsilon$ and Model III: $Y = A + BX_1 + CX_2 + \varepsilon$. It can be very difficult to determine which model (if any) is the "correct" one. Two contradictory rules apply:
   A. The more parameters, the better the fit. In this context, Model III is to be preferred to Model I.
   B. The simpler, more straightforward model is more likely to be correct. In this context, Model I is to be preferred to Model III.
5. Goodness of fit is not prediction. The goodness-of-fit criteria used to estimate the values of model parameters minimize the sum of squares for the historical data $\Sigma(Y_{observed} - Y_{model})^2$. But minimizing this sum of squares is no guarantee that when we continue to gather data, we will minimize the sum $\Sigma(Y_{observed} - Y_{predicted})^2$ based on the square of the difference between what we observe in the future and what our model predicts. If the intent is to, say, predict future loss of earnings, this distinction can be critical.
6. If the independent variables are categorical, a k-sample comparison, an analysis of variance, or some data mining technique may be more appropriate.[42]

---

[41] See, for example, *Chlorine Chemistry Council v. EPA*, (D.C. Cir. 2000).
[42] See Chapter 10.

Extrapolation versus Interpolation

The nonzero value of b in the preceding example is troubling. Are we really to believe that in the year 0, many hundreds of years before the wholesaler was in business, its earnings were $27,520? The answer is no. Our estimation procedures make sense over the range of values studied. The error lies in trying to extrapolate that straight line outside of that range.

Here is another example. In 1981, the Puerto Rico Maritime Shipping Authority, a common carrier, filed for a rate increase based primarily on an expectation of rising fuel prices during 1981. During 1980, prices had risen steadily each quarter, from $19.60 per barrel to $20.60, $21.39, and $28.03. After fitting a regression model to this data, the carrier asked for similar increases in its rates for 1981 and beyond. The government did not grant the increase. Oil prices stabilized in 1981 and remained stable for the next three or four years. Today, with the advantage of hindsight, we know the government was right in resisting the rate change.[43]

Not Always Evident to the Naked Eye

The presence of a strong relationship between two variables may not always be evident to the naked eye.

> The District Court also pointed to "scattergrams" or graphs based on the data in respondents' regressions, concluding that these graphs displayed the salaries of blacks and whites "in a completely random distribution." Yet, as pointed out by the U.S. ... the very purpose of a regression analysis is to organize and explain data that may appear to be random. See *Fisher* [1980]. Thus, it is simply wrong to give weight to a scattergram while ignoring the underlying regression analysis. Respondents' strategy at trial was to declare simply that many factors go into making up an individual employee's salary; they made no attempt that we are aware of — statistical or otherwise — to demonstrate that when these factors were properly organized and accounted for there was no significant disparity between the salaries of blacks and whites.[44]

See also Anscombe [1973].

---

[43] *Puerto Rico Maritime Shipping Authority v. Federal Maritime Comm.*, 678 F.2d. 327, 337-342 (D.C. Cir. 1982).

[44] *Bazemore v. Friday*, 478 U.S. 385, 404, fn. 14. This case is discussed at length in Section 12.5.

11.3.2 Comparing Two Populations

Prior to 1990, no Hispanic had ever been elected a supervisor in Los Angeles County despite the fact that Hispanics comprised 28% of the total population and 15% of the voting-age citizens. In a suit[45] to compel redistricting to create a majority Hispanic district, the plaintiffs offered in evidence two regression equations to demonstrate differences in the voting behavior of Hispanics and non-Hispanics:

$$Y_{hi} = C_h + b_h X_{hi} + \varepsilon_{hi}$$

$$Y_{ti} = C_t + b_t X_{hi} + \varepsilon_{ti}$$

where Y_{hi}, Y_{ti} are the predicted proportions of voters in the ith precinct for the Hispanic candidate and for all candidates, respectively; C_h, C_t are the percentages of non-Hispanic voters who voted for the Hispanic candidate and all candidates; b_h, b_t are the added percentages of Hispanic voters who voted for the Hispanic candidate and all candidates; X_{hi} is the percentage of registered voters in the ith precinct who are Hispanic; and ε_{hi}, ε_{ti} are random or otherwise unexplained fluctuations.

If there are no differences in the voting behavior of Hispanics and non-Hispanics, then we would expect our estimates of b_h, b_t to be close to zero. Instead, the plaintiffs showed that the best fit to the data was provided by the equations:

$$Y_h = 7.4\% + .110\ X_h$$

$$Y_t = 42.5\% - .048\ X_h$$

Of course, other estimates of the Cs and bs are possible, as only the Xs and Ys are known with certainty; it is conceivable, but unlikely, that few if any Hispanics voted for the Hispanic candidate. Moreover, in this case, as in *NAACP v. City Of Niagara Falls*, "Simple regression does not allow for the effects of racial differences in voter turnout; it assumes that turnout rates between racial groups are the same."[46]

---

[45] *Garza et al. v. County of Los Angeles*, 918 F.2d 763 (9th Cir.), <u>cert. denied</u>, 498 U.S. 1028 (1991).

[46] 65 F.3d 1002, n. 2 (2nd Cir. 1994).

A Matter of Interpretation

From the plaintiff's point of view, the Hispanic candidate received on the average 7.4% of the votes of non-Hispanics and 18.4% of the votes of Hispanics. The implication is that all Hispanics, rich or poor, old-established families or recently arrived immigrants, behaved in similar fashion. The defendants argued that the perceived differences could be attributed to differing economic and social conditions among precincts or neighborhoods and that within each precinct (where the housing and the housed were relatively homogeneous) all the people voted more or less the same way regardless of their ethnic origins. This issue continues to be a matter of strong controversy.[47]

11.4 Summary

Two variables are said to be correlated if a change in one appears to be accompanied by a change in the other. The correlation coefficient R takes values between -1 and $+1$. If $R = +1$, the variables are totally dependent, rising and falling almost as one. If $R = -1$, the variables are also totally dependent; however, when one variable falls, the other rises. If $R = 0$, the variables are completely independent.

Often, when two variables are related, one is the cause (the independent variable) and the other is the effect (the dependent variable). In this situation, R^2 provides a measure of the variation in the dependent variable that is explained by the second, independent variable.

If $R = 0$, there may exist a relationship between the two variables of the form $Y = b_0 + b_1 f[X] + \varepsilon$, a linear regression of the dependent variable Y upon the independent variable X. In this context, R^2 provides a measure of the credence to give this equation for the purpose of predicting Y given X.

Six limitations to the regression method were listed in the sidebar in this chapter. In the next chapter, we will consider the need to consider the influences of multiple variables before drawing a conclusion.

---

[47] See Rubinfeld [1991]; Klein and Freedman [1993, p. 38]; *Thornburg v. Gingles*, 478 U.S. 30, 53, n. 20 (1986); *Teague v. Attala County*, 92 F.3d 283, 285 (5th Cir. 1996); *Houston v. Lafayette County*, 56 F.3d 606, 612 (5th Cir. 1995); *Lewis v. Alamance County*, 99 F.3d 600, 604, n. 3 (4th Cir. 1996); *Aldasoro v. Kennerson*, 922 F. Supp. 339 (S.D. Cal. 1995); *Johnson v. Miller*, 864 F. Supp. 1354, 1390 (S.D. Ga. 1994), aff'd, 515 U.S. 900 (1995); *Romero v. City of Pomona*, 665 F. Supp. 853, 860 (C.D. Cal. 1987), aff'd, 883 F.2d 1418 (9th Cir. 1989).

Chapter 12

Multiple Regression

Statistical analysis is perhaps the prime example of those areas of technical wilderness into which judicial expeditions are best limited to ascertaining the lay of the land.[1]

There is some argumentation in the briefs about the relative merits of multiple linear regression and logistic fitting analysis. Neither side, however, either in the briefs or at the oral argument, bothered to explain, in intelligible terms or otherwise, what these terms mean.[2]

In Chapter 10, we saw that the courts distrust arguments predicated on the presence or absence of a single factor. In the last chapter, we noted the improvements that could be obtained by taking advantage of the detailed relationships among variables via regression analysis. In this chapter, we consider regression methods for assessing the relative contributions made by multiple factors in cases involving earnings, housing, and employment. We also consider actual and potential counter arguments. Concepts introduced include collinear and partially dependent variables; goodness-of-fit; linear, nonlinear, and logistic regression; and cohort analysis.

---

[1] *Appalachian Power Co. v. EPA*, 135 F.3d 791 (D.C. Cir. 1998).
[2] *Craig v. Minnesota State University Board*, 731 F.2d 465, 476, fn. 14 (8th Cir. 1984).

12.1 Lost Earnings

Very few cases involve only one explanatory variable. Suppose, in the example considered in Section 11.3.1, the supplier's attorney had responded that while the wholesaler and similarly situated companies may have done well in the pre-conspiracy period, the entire industry subsequently suffered reverses. Consequently, the attorney proposes the following formula for earnings determination in which industry sales are incorporated:

Earnings = $27,520 + $11.7 * Year + 0.0014 * $Industry_Sales

As before, *earnings* is the dependent variable, and *year* and *industry sales* are the independent variables. Note that the coefficient of the explanatory variable *Years* has decreased from $12.5 per year as shown in Section 11.3.1 to $11.7 per year as a result of the addition of *industry sales* to the equation.

The fact that the coefficient of industry sales is several orders of magnitude smaller than the year coefficient does not mean it is less important as a predictor or explains less of the variation in earnings than the year coefficient does. These coefficients also include scaling factors that depend upon (1) the units in which the various variables are measured[3] and (2) the intrinsic variability of each variable relative to the variability of all the other variables. Only the individual contributions of each variable to R (discussed below) can provide insight into their relative importance.

In the most general form of a linear regression, we write the dependent variable Y as a linear function of n explanatory variables:

$$Y = b_0 + b_1 f_1[X_1] + b_2 f_2[X_2] + \dots + b_n f_n[X_n]$$

where b_0, b_1, ... b_n are constants to be determined; X_1, X_2, ... X_n are explanatory or predictor variables such as year and industry sales; and f_1, f_2, ... f_n are nondecreasing functions of these variables.

As in the previous chapter, we will want to tell the court what proportion R^2 of the variable Y is explained by X_1, X_2, ... X_n. Let y_1, ... y_n denote the actual observations, and y_1, ... y_n the values we would predict using the regression equation:

$$R^2 = \frac{Var[y] - \frac{1}{n}\sum_{i=1}^{n}(y_i - y_i)}{Var[y]}$$

---

[3] For example, if we measured time in months rather than years, the coefficient would be 11.7/12 = 0.975.

that is, R^2 measures the proportionate reduction in the variability of Y as a result of our knowledge of the explanatory variables X_1, X_2, ... X_n.

In ordinary least squares (OLS), the most common regression method, the coefficients b_0, b_1, ... b_n are chosen so as to minimize the sum of the squared deviations between the observed and the predicted values, that is, to minimize $\sum_{i=1}^{n} (y_i - y_i)^2$, and thus, to maximize R^2. A dozen or so off-the-shelf computer programs will do the computations and provide estimates of the individual contributions made by each of the explanatory variables.

12.2   Multiple Applications

A regression analysis can serve multiple ends as this next example illustrates. The Appalachian Power Company appealed an EPA ruling that would have set new nitrogen oxide (NOx) emission limits for some of its existing boilers.[4] The EPA's rationale was that improvements in technology made it possible for power companies to meet higher standards. Accordingly, it proposed that the company retrofit its coal-fired boilers with new emission control devices. The company objected to the cost.

> EPA was permitted to revise the Group 1 limits by January 1, 1997, to apply to Phase II boilers if it determined that "more effective low NOx burner technology [was] available."[5]

> EPA revised the Group 1 limits after determining, as required by section 407(b)(2), that boilers with low NOx burners were achieving lower emission levels than the limits promulgated in 1995 and therefore that more effective low NOx burner technology was available. [This determination was the result of a regression analysis in which EPA constructed equations capturing the reductions achieved by Group 1, Phase I boilers and applied these equations to the uncontrolled emission rates of Group 1, Phase II boilers.]

> We thus think it reasonable, as a preliminary matter, for EPA to find that "more effective low NOx burner technology" exists if improved performance for already existing burners can be shown.

---

[4] *Appalachian Power Co. v. EPA*, 135 F.3d 791 (D.C. Cir. 1998).
[5] 42 U.S.C. §7651f(b)(2).

Appalachian Power next argues that even if "more effective low NOx burner technology" is given the meaning we approve today, EPA has failed to show that boiler performance has improved. EPA asserts that its regression analysis shows that boilers retrofitted with low NOx burners can achieve lower emission levels than the limits deemed adequate by the 1995 rule.

Because Appalachian Power's challenge, although framed in the most general of terms, is at root a challenge to EPA's analytical model, we must consider whether the use of that model was arbitrary and capricious.[6] Our analysis is guided by the deference traditionally given to agency expertise, particularly when dealing with a statutory scheme as unwieldy and science-driven as the Clean Air Act. As we have previously noted, so long as EPA "acted within its delegated statutory authority, considered all of the relevant factors, and demonstrated a reasonable connection between the facts on the record and its decision," we will not interfere with its conclusion.[7]

Statistical analysis is perhaps the prime example of those areas of technical wilderness into which judicial expeditions are best limited to ascertaining the lay of the land. Although computer models are "a useful and often essential tool for performing the Herculean labors Congress imposed on EPA in the Clean Air Act,"[8] their scientific nature does not easily lend itself to judicial review. Our consideration of EPA's use of a regression analysis in this case must therefore comport with the deference traditionally given to an agency when reviewing a scientific analysis within its area of expertise without abdicating our duty to ensure that the application of this model was not arbitrary. As we have noted, it is only when the model bears no rational relationship to the characteristics of the data to which it is applied that we will hold that the use of the model was arbitrary and capricious.[9] Therefore, while we will examine each step of EPA's analysis to satisfy ourselves that the agency has not departed from a rational course, we will not take it upon ourselves, as nonstatisticians, to

---

[6] See 42 U.S.C. §7607(d)(9) (1994) authorizing reversal of actions under the Act found to be "arbitrary, capricious, an abuse of discretion, or otherwise not in accordance with law."

[7] *Ethyl Corp. v. EPA*, 51 F.3d 1053, 1064 (D.C. Cir. 1995).

[8] *Sierra Club v. Costle*, 657 F.2d 298, 332 (1981).

[9] See *American Iron & Steel Inst. v. EPA*, 115 F.3d 979, 1005 (D.C. Cir. 1997); *Chemical Mfrs. Ass'n. v. EPA*, 28 F.3d 1259, 1265 (D.C. Cir. 1994).

perform our own statistical analysis — a job more properly left to the agency to which it was delegated.[10]

EPA's determination of the revised Group 1 emission rates involved four steps: (1) the construction of a database consisting of Group 1, Phase I boilers already employing low NOx burner technology; (2) the derivation of two equations (one each for tangentially fired boilers and wall-fired boilers) that captured the percent reduction in emissions from the uncontrolled emission rates achieved by the boilers in the database; (3) the application of these equations to the uncontrolled emission rates of Group 1, Phase II boilers; and (4) the setting of emission rates for Group 1, Phase II boilers based on the range of data resulting from the application of the equations. We examine each of these steps in turn.

12.2.1 Construction of the Database

EPA began its determination of whether the Group 1 limits should be revised by constructing a computerized database consisting of all known boilers that had installed only low NOx burners 15 subsequent to November 15, 1990 (the date the amendments to the Act were enacted), and for which there existed at least 52 days of data measured by continuous emission monitors (CEMs). This database consisted initially of 24 wall-fired boilers and 9 tangentially fired boilers. In response to the recommendations of several commenters that various boilers be included in or excluded from this database, EPA formalized and expanded its selection criteria into Data Quality Objectives (DQOs) "rigorous and precisely defined rule tables" used to select candidates for the database. Application of the DQOs resulted in a new database consisting of 39 wall-fired boilers and 14 tangentially fired boilers, a result that EPA believed would "increase the overall representativeness of the database."[11]

EPA then considered the lowest average NOx emission rate each boiler in the database had sustained for at least 52 days when CEM data were available (the low NOx period). To take

---

[10] As we saw in Section 3.2.2 and will see in many of the examples later in this chapter, the best defense is to provide an alternate statistical model.

[11] Citations to Federal Regulations are omitted throughout.

into account the fact that the emissions rate immediately after low NOx burner installation might not be representative of a boiler's emissions rate at full operating capacity, EPA also analyzed emission rates for a time period beginning 30 days after resumption of operation after installation until the end of the available CEM data as well as for a time period beginning with the first hour of the low NOx period until the end of the available CEM data. In response to comments that suggested that the 52-day period alone was insufficient to determine actual emission rates, EPA selected for the final rule the time period beginning with the first hour of the low NOx period until the end of the available CEM data (the post-optimization period) as the basis for assessing low NOx burner performance.

Moreover, we can find no apparent defects in the database itself. In constructing the database for the final rule, EPA applied the DQOs not only to those boilers used in the proposed rule analysis but also to those boilers that commenters requested that EPA consider as well as to additional boilers identified by EPA as using low NOx burner technology. This resulted in the addition of 20 boilers to the database (which ultimately contained a total of 39 wall-fired boilers and 14 tangentially fired boilers).

Although the relatively small number of tangentially fired boilers might be cause for looking more closely at the regression analysis for this subgroup,[12] Appalachian Power has not controverted EPA's assertion that the database is representative of the entire boiler population.

In this respect, EPA has identified all likely candidates for the boiler database as well as been responsive to comments. Appalachian Power's assertion that the emission rates reflect boilers employing beyond-burner technology is not supported by the record.

12.2.2 Construction of the Equations

Using the database, EPA constructed two linear regression equations, one for wall-fired boilers and one for tangentially fired

---

[12] See Rubinfeld [1994] noting that 30 data points are typically seen as sufficient for regressions with a small number of explanatory variables.

boilers, that captured the percent reduction in emissions with low NOx burner technology as a function of the uncontrolled emission rate. As EPA noted in the preamble to the final rule, the use of a regression model rather than a simple extrapolation from the low NOx burner database would enable EPA better to predict the effect of installing low NOx burner technology on Phase II boilers.

It is commonly understood that the more variables that are included in a regression analysis, the more likely it is that the model describes accurately the phenomenon it is being used to explain. As the Supreme Court has noted in the employment discrimination context, "the omission of variables from a regression analysis may render the analysis less probative than it otherwise might be," but it does not render the analysis completely devoid of value.[13] Nevertheless, a number of commenters, Appalachian Power among them, argued that EPA's analysis failed to take into account several operational factors associated with low NOx burners, including normal aging and wear of equipment, increased particulate emissions, auxiliary equipment design, and furnace configuration, all of which arguably could have an effect on the level of NOx emissions. EPA responded to this concern by using the post-optimization period rather than the 52-day period for analysis, which it believed would "reasonably account for variation in operating conditions among Group 1 boilers. The claim that there are various problems due to aging of equipment that have not yet been encountered," the agency continued, "is speculative and unsupported."

While EPA's response could have been more extensive, it does not suggest that the agency's use of the regression models was arbitrary and capricious. As we have previously noted, the party challenging the use of such a model "cannot undermine a regression analysis simply by pointing to variables not taken into account that might conceivably have pulled the analysis's sting."[14] Rather, that party must identify clearly major variables, the omission of which renders the analysis suspect. This conclusion, derived from employment discrimination cases, holds

---

[13] *Bazemore v. Friday,* 478 U.S. 385, 400 (1986).

[14] *Koger v. Reno,* 98 F.3d 631, 637 (D.C. Cir. 1996) (dicta). See also *Segar v. Smith,* 738 F.2d 1249, 1277 (D.C. Cir. 1984) noting that where there is no reason to conclude that the omitted variable correlates with the dependent variable, the omission will not affect the validity of the analysis.

equally true in this context, even more so because of the
deference due to an agency's scientific analysis. Neither the
commenters before EPA nor Appalachian Power here before us
has offered any data to support the assertion that additional
factors not accounted for in EPA's regression analysis would
have a significant effect on NOx emissions. The regression
equations were constructed based on the data available or
reasonably predictable at the time of the final rule; to require
EPA to take into account variables for which no data existed
would be to require that it engage in precisely the type of
arbitrary rulemaking the Act forbids.

12.2.3 Application of the Equations

The next step of EPA's analysis was to calculate, through the
application of the regression equations developed to the uncontrolled rates of the Phase II boilers, the NOx emission rate each
Phase II boiler could be expected to achieve through a low
NOx burner retrofit. Appalachian Power's primary challenge to
this step of the analysis is that the results generated by the
regression equation are faulty because they do not include the
uncertainty inherent in the calculation — in other words, the
true reduction in NOx emissions associated with a particular
retrofit might be somewhat greater than or less than the amount
yielded by the regression equation. As a result, Appalachian
Power contends, because EPA based its revised emission limits
on what represents the midpoint between the uncertainty
boundaries, the predicted emission levels are based on levels
achievable by only 50 percent of the Phase II boilers.

Although Appalachian Power's assertion that the results are
subject to some uncertainty is correct, we do not believe its
complaint constitutes a telling critique of EPA's analysis. In any
regression analysis, the line described by the regression equation represents the best possible fit to the data; some points
will necessarily be plotted above the line and some will fall
below the line (except in the rare circumstance in which the
line is a perfect fit to the data). While each data point will be
associated with some residual (the difference between actual
and fitted values), so long as this residual is within acceptable
statistical limits, the fact that some data points necessarily fall
below the line does not render the regression analysis invalid.

As we have noted in similar circumstances, "[t]hat the model does not fit every application perfectly is no criticism; a model is meant to simplify reality in order to make it tractable."[15] To invalidate a model simply because it does not perfectly fit every data point "would be to defeat the purpose of using a model."[16] Appalachian Power does not suggest in its argument before us that the uncertainty surrounding the data points is statistically unacceptable, only that it exists. We would not deem that sufficient to label EPA's model arbitrary and capricious.

While "[a]s a general rule, courts should be reluctant to rely solely on a statistic such as R to choose one model over another,"[17] it cannot be said from these values that EPA's use of this statistical model represented unreasoned decisionmaking.

This is not, certainly, like the case in *Sierra Club*, in which we rejected a 92 percent sulfur removal rate that was based solely on evidence that "only one commercial scale plant and one small pilot unit can almost but not quite meet the standard."[18] In this case, 23 of 39 wall-fired boilers and 9 of 14 tangentially fired boilers in Group 1, Phase I can currently meet the revised limits.[19] Because the statute requires only a determination that more effective low NOx burner technology is "available" for a class of boilers, the fact that, as Appalachian Power claims, some individual boilers cannot currently meet the revised limits does not serve to invalidate the rule.

12.2.4 Determination of the Limitation

Finally, EPA used the rates resulting from the regression equations to determine "reasonably achievable emission limitation[s]" for Phase II boilers.[20] Appalachian Power asserts, however, that the predicted controlled emission rates of many boilers are so close to the emissions limits that any error in the prediction would render these boilers in violation of the limits. In addition, it claims that many utilities typically attempt to overcontrol

---

[15] *Chemical Mfrs. Ass'n. v. EPA*, 28 F.3d 1259, 1264 (D.C. Cir. 994).
[16] Id. at 1265.
[17] Rubinfeld, supra, at 457; see also *Segar*, 738 F.2d at 1282, n. 27.
[18] 657 F.2d at 363.
[19] 61 Fed. Reg. at 67,123-67,124 (Tables 4 and 5).
[20] 61 Fed. Reg. at 67,130.

emissions so that any fine tuning of the boiler will not bring the boiler over the emission limit. If the rule deems this "over-controlled" emissions level achievable, Appalachian Power claims, utilities will be penalized for anticipating control diffi-culties. For both these reasons, Appalachian Power argues that EPA should have included a compliance margin in the NOx emission limits.

Again, we find this challenge insufficient to vacate the rule. The first reason that this is so is a statutory one: The Act permits EPA to revise the emissions limitations upon a finding that "more effective low NOx burner technology is available."[21] The fact that these boilers can achieve lower emission levels with low NOx burner technology — even if they depend on a cushion between those levels and the emissions limits — demonstrates that the statutory requirement has been satisfied. Moreover, as EPA has noted in the preamble to the final rule, boiler owners who fear that tuning may send them over the allowable limits may use the alternative emission limitations (AEL) and averaging options provided in the Act to ensure that their total NOx emissions from all affected units comply with EPA regulations. Finally, we note, as EPA has pointed out, that the limit applies to a unit's average annual emission rate rather than to a monthly or a daily emission rate. This means that a boiler may overemit on some days and underemit on others and still be deemed in compliance with its emission limit. Given these various options, there is no reason that EPA's failure to build a compliance margin into the limits themselves should render them arbitrary and capricious. We therefore reject Appalachian Power's chal-lenge on this front, as with its other substantive challenges to the Group 1, Phase II limits.

12.3 Collinearity and Partial Correlation

Multiple linear regression entails a further, major complication not present in simple linear regression, that is best illustrated by the following example. Suppose a not-very-helpful executive at Spray-Rite proposes adding phalange-drawback-turnover[22] to the equation in Section 12.1. What he

---

[21] 42 U.S.C. §7651f(b)(2).
[22] Yes, I made up that name.

really does is simply slip industry sales in under another name, so that in the equation:

$$Y = b_o + b_1 Y + b_2 I + b_3 P$$

the letters I and P really stand for the same variable. As a result, the coefficients b_2, b_3 could take many different values and still be correct, providing that $b_2 + b_3 = 0.0014$, the coefficient of industry sales in the original equation.

A total dependence of this sort among the so-called independent variables, termed *collinearity*, is rare, but not unknown. In its most general form, the n+1st independent variable introduced into a model proves to be a linear function of the previous n, that is, $X_{n+1} = a_1 X_1 + a_2 X_2 + \ldots + a_n X_n$ where at least one of the constants a_1 is not equal to zero. As a result, the coefficients could take many different values and we would have difficulty establishing which (if any) of the variables made the greatest contribution to R^2. To avoid the difficulties associated with collinearity, add a further variable to a model only after making a very careful comparison with the variables that are already present.[23]

While collinearity is rare, few variables are completely independent of one another. Consider intelligence, education, and income. When dependent variables are included in a regression equation, the values of the coefficients will depend upon the order in which the variables are entered into the model. In consequence, one cannot readily conclude, for example, whether education or intelligence has the greater impact on income.

The good news is that this problem has a solution; the bad news is that it has several of them. In the *forwards* method, a variable at a time is introduced into the model and only the variables yielding the highest values of R² are retained at each step. In the *backwards* method, one starts with all the variables at once in any order and then the variable making the least contribution to R^2 is removed at each step.[24] Still other regression methods go forwards, then backwards, and so forth. The results may differ from method to method. At least one court has rejected a regression analysis because of the method that was employed.[25]

One way to demonstrate to the court that a specific variable is important for determining Y is to compare the values of R^2 before and after this

---

[23] While a statistician may be the ideal expert to consult on which statistic or which of the many methods of regression or data mining to use, domain specialists (experts on boilers, or discrimination, or DNA analysis) are the ones who should be paired with the statistician, both on the witness stand and in the laboratory, when deciding which variables to utilize and in what order.

[24] Naturally, a computer does all the thousands of computations.

[25] *Eastland v. Tennessee Valley Authority*, 704 F.2d 613 (11th Cir. 1983).

variable is included among the explanatory variables. Another, better way, is to provide adequate collateral evidence for your choices including the results of other studies and the testimony of domain experts (biologists, economists, and sociologists).

The best advice is to lay out all the details of your model *before* you collect the data.

> In multiple regression analysis, one builds a theoretical statistical model of reality and then attempts to control [account] for all independent variables which [while] measuring the effect of the variable of interest upon the dependent variables. Thus a properly done study begins with a decent theoretical idea of what variables are likely to be important.[26]

12.4 Defenses

Because of the complexity of the multiple regression method, it can be attacked on a variety of grounds:

- The data is incomplete or inaccurate.
- Essential variables are omitted.
- Tainted variables are included.
- Distinct groups are wrongly aggregated in a single regression.
- The model is not unique.
- The model is a poor predictor and thus inadequate or incorrect.
- The wrong methodology is used to derive the coefficients.
- The regression assumptions are not satisfied.

Many of these criticisms were echoed in the district court's opinion in the landmark case of *McCleskey v. Zant*.[27] The plaintiff, an African-American, appealed his death sentence as discriminatory. To demonstrate that sentencing decisions in Georgia were affected by the races of both the defendants and the victims, David Baldus and his colleagues studied 2000 murder cases that occurred in Georgia during the 1970s, taking account of 230 variables that could have explained the disparities on nonracial grounds. Still, the district court felt that in many respects the Baldus data was incomplete:

---

[26] *McCleskey v. Zant*, 580 F. Supp. 338, 353 (1984).
[27] Ibid.

- The questionnaires used to obtain the data failed to capture the full degree of the aggravating or mitigating circumstances.
- The researchers could not discover whether penalty trials were held in many of the cases, thus undercutting the value of the study's statistics as to prosecutorial decisions.
- In certain cases, the study lacked information on the races of the victims in cases involving multiple victims, on whether the prosecutor offered a plea bargain, and on credibility problems with witnesses.

The district court also objected to several aspects of the analysis:

- The assumption that all of the information available from the questionnaires was available to the juries and prosecutors when the case was tried.
- The instability of the various models. Even with the 230-variable model, consideration of 20 further variables caused a significant drop in the statistical significance of race.
- The high correlation between race and many nonracial variables
- The inability of any of the models to predict the outcomes of actual cases.

The Eleventh Circuit appeals court did not buy into these criticisms and accepted the Baldus study as valid but insufficient.[28] Judge Johnson went further in his dissent:

- The flaws in the database were not significant; some apparent mismatches were the results of improvement in coding technique.
- Collinearity was not a relevant concern since it acted to reduce the statistical significance of the variables of interest.
- Adding more variables produced collinearity and thus reduced the effects of race.
- Like the Baldus study, the best models use the fewest variables in the most effective fashion.

12.4.1 Failure to Include Relevant Factors

In *Smith v. VCU*,[29] the plaintiffs-appellants were five male professors at Virginia Commonwealth University (VCU) who filed claims under the Equal Pay Act and Title VII of the Civil Rights Act of 1964 in the Eastern District of Virginia. The appellants objected to pay raises that VCU gave to female faculty members but not to males.

---

[28] *McCleskey v. Kemp*, 753 F.2d 877, 896 (11th Cir. 1985).
[29] 84 F.3d 672 (4th Cir. 1996).

In the spring of 1988, VCU appointed a committee to evaluate its pay structure to determine whether female professors were victims of sex-based discrimination in pay. The committee employed a multiple regression analysis, which compares many characteristics within a particular set of data and enables the determination of how one set of factors is related to another, single, factor. The VCU study controlled for such differences as doctoral degree, academic rank, tenure status, number of years of VCU experience, and number of years of prior academic experience. Any difference in salary after controlling for these factors was attributed to sex. The study included only tenured or tenure-eligible instructional faculty at the rank of assistant professor or higher.

> The first regression study in the summer of 1989 showed a $1354 difference in salaries not attributable to permissible factors. A second analysis run in the summer of 1991 showed a difference of $1982. Until the study, the compensation system at VCU had been based on merit alone. A professor was awarded a pay increase after a detailed annual review, provided funds were available. Merit factors considered in the annual review were teaching load, teaching quality, quantity and quality of publications, and service to the community (the "performance factors"). The department chair recommended a pay raise to the dean, and the dean awarded a pay raise, subject to approval from VCU's Board of Visitors.

> Salaries vary widely from department to department. The multiple regression analysis did not include the performance factors because VCU contended that these would be too difficult to quantify. VCU maintained that indirect performance variables were already included in the study in the form of academic rank, status, and experience. The study also did not take into account a faculty member's prior service as an administrator. Administrators are paid higher wages, and faculty members retain this increase in salary when they return to teaching, thus inflating faculty salaries. Most of the faculty that had previously served as administrators were men. Furthermore, the study did not include career interruptions when measuring academic experience. Finally, the Study Committee worked under the assumption that there was no reason to suspect that female faculty members were less productive on the average than male faculty members. After the study was completed, VCU approved more than $440,000 in funding to increase female faculty salaries. These funds were outside of the normal salary process.

The pay increases were implemented by the Salary Equity Implementation Committee made up of three women. Female faculty members had to apply for a pay increase by submitting a curriculum vitae or a narrative statement and a vitae. Of the 201 women eligible for salary review, 172 requested it. All women who requested a review received an increase in salary.

After the district court's denial of the plaintiff's motion for summary judgment and the grant of VCU's motion, the plaintiffs filed a motion to alter or amend the judgment. In this motion, the plaintiffs offered the affidavit of expert witness Dr. Fred McChesney. McChesney contended that the performance factors VCU claimed it could not quantify had in fact been included in several studies of various faculty systems, and that the inclusion of the performance factors and other variables was necessary to ensure accurate statistical data. McChesney also contended that there was data to dispute VCU's assumption that women were as equally productive as men. In response, VCU's expert witness, Dr. Rebecca Klemm, stated that she ran several various statistical studies with VCU's raw data and found a salary gap to be consistent with that found in the study. McChesney never conducted a pay study himself. The district court denied the motion to alter or amend the judgment.

We agree that there is a material question of fact as to whether there was a manifest imbalance in compensation between the male and female faculty. We do not reach the question of whether the plan unnecessarily trammeled the rights of the male faculty. VCU relied on its multiple regression analysis in determining that there was a manifest imbalance and in instituting its affirmative action pay raises. *Therefore, the validity of the plan stands or falls on the soundness and accuracy of the regression study.* The appellants contend that the multiple regression study was flawed because of the inclusion of an inflated pool of faculty members, and the failure to account for significant variables that could have a bearing on wage differences between the male and female faculty members.

Population for Comparison Purposes

Statistics are often used to determine whether a manifest imbalance exists. In *Johnson*,[30] the Court stressed that valid statistical analyses must include logical comparisons. The Court stated:

---

[30] *Johnson v. Transportation Agency*, 480 U.S. 616, 631-632 (1987).

in determining whether an imbalance exists that would justify taking sex into account, a comparison of the percentage of minorities or women in the employer's work force with the percentage in the area labor market or general population is appropriate in analyzing jobs that require no special expertise .... Where a job requires special training, however, the comparison should be with those in the labor force who possess the relevant qualifications.[31] Because the instant case does not involve a simple question of whether women are underrepresented as compared to the available labor pool in a particular job, a more sophisticated statistical analysis was necessary to determine if female faculty members were paid less than male faculty members based solely on their sex. It is still necessary, however, to avoid illogical comparisons. The appellants contend that the study used illogical comparisons by comparing an inflated pool of faculty members. The study included male faculty members who had returned from higher paying positions in the VCU administration, but it did not account for this salary differential. An inflated pool can undermine the validity of a statistical study to determine imbalances.[32] Appellants' expert, Dr. McChesney, stated that failure to include a faculty member's status as a former administrator could easily have caused a salary differential that was not attributable to sex. Dr. Henry, the man who designed VCU's regression study, stated that inclusion of this factor in the study would have had an effect on the study, and that if he had had the information, he would have included it. The appellants clearly produced evidence to support a finding that the pool was inflated. More importantly, VCU did not include major factors that are legitimate reasons for wage disparity in its multiple regression analysis. Although failure to include some factors affects a study's weight, not its admissibility, a study must include all major factors.[33]

In brief, the VCU study:

- Failed to account for a professor's status as a former administrator.
- Failed to measure the amount of time actually spent in teaching instead of the lapse of time since a professor began teaching.
- Most significantly, failed to include the performance factors.

---

[31] Id. at 632.

[32] Id. at 636; see also *EEOC v. Sears, Roebuck & Co.*, 839 F.2d 302, 322-324 (7th Cir. 1988).

[33] *Bazemore v. Friday*, 478 U.S. 385, 400 (1986).

VCU maintains that it could not have included the performance factors because, due to their subjective nature, they are not suitable for statistical analysis. The appellants' expert, nevertheless, stated that the study was not valid without adding the performance factors, and that studies performed by disinterested outside researchers "have regularly included productivity measures such as teaching loads and publications," and these studies have shown that productivity has a positive effect on the level of faculty compensation.

Because VCU's pay system is based on merit, and the factors on which faculty pay scales are decided are the very performance factors left out of the regression analysis, *it cannot be said that VCU's study included all major factors.* The inclusion of the performance factors could very well alter the results of the multiple regression analysis, and there is a dispute of material fact as to whether inclusion of these factors is feasible. Given the number of important variables omitted from the multiple regression analysis, and the evidence presented by the appellants that these variables are crucial, a dispute of material fact remains as to the validity of the study. Therefore, the decision of the district court granting VCU's motion for summary judgment is REVERSED.[34]

12.4.2 *Negligible Predictive Power*

In *EEOC v. IBM Corp,* [35] the Equal Employment Opportunities Commission (EEOC) sued IBM on behalf of an African-American who alleged discrimination by IBM on the basis of race with respect to performance plans, evaluations, promotions, and pay. The EEOC's expert developed an equation that would allow him to predict monthly salaries of white employees for each of four groups based on job function. Once developed and tested, this equation was applied to black employees who had the same time in level. He also conducted a separate multiple regression analysis using race as a dummy variable. The court faulted these multiple regression studies on the following grounds:[36]

[34] Id. at 1226.
[35] 583 F. Supp. 875 (Md. 1984).
[36] Id. at 897-899.

1. Employee groupings included a substantial number of positions that the court previously had ruled were not encompassed by the lawsuit.
2. The studies did not separate out employees who were (1) with another division of IBM, (2) employed in another state, or (3) in a job category other than managerial or professional.
3. The studies failed to take into account a critical factor: seniority.
4. They failed to take into account trend data that showed that differences in seniority, salary levels, and salary amounts diminished over time.
5. Explanatory variables were insufficient. Only current appraisal rating and time in current position were used by the EEOC while IBM's statistical expert factored in salary level, seniority, time in level, current appraisal, and education.
6. The predictive power of the regression equations, yielding an R^2 less than 0.02, was negligible.

12.4.3 Validation

How can one be sure that a model fairly represents the underlying cause and effect mechanisms? Goodness of fit is not prediction and a high R^2 can be meaningless for the reasons cited above and in Sections 8 and 11.3. One of the few satisfactory solutions is to divide the data into two parts; fit a model to the first part of the data only; and, after all model parameters have been determined, attempt to fit that same model to the second part of the data. If the model fits, use it. Otherwise, discard it and begin again. The final model may have a lower R^2 than the one adopted initially, but it will be less open to criticism.

12.5 Rebuttal Decisions

Failure to include variables will affect the probativeness of an analysis, not its admissibility. Collateral evidence in conjunction with statistical evidence can be persuasive. Here is an example. *Bazemore v. Friday*,[37] reviewed allegations of racial discrimination in employment and in provision of services by the North Carolina Agricultural Extension Service.

> The first issue we must decide is whether the Court of Appeals erred in upholding the District Court's finding that petitioners had not proved by a preponderance of the evidence that respondents

[37] 478 U.S. 385 (1986).

had discriminated against black Extension Service employees in violation of Title VII by paying them less than whites employed in the same positions. The Court of Appeals reasoned that the Extension Service was under no obligation to eliminate any salary disparity between blacks and whites that had its origin prior to 1972 when Title VII became applicable to public employers such as the Extension Service. It also reasoned that factors, other than those included in petitioners' multiple regression analyses, affected salary, and that therefore those regression analyses were incapable of sustaining a finding in favor of petitioners.[38]

The error of the Court of Appeals with respect to salary disparities created prior to 1972 and perpetuated thereafter is too obvious to warrant extended discussion: that the Extension Service discriminated with respect to salaries prior to the time it was covered by Title VII does not excuse perpetuating that discrimination after the Extension Service became covered by Title VII. To hold otherwise would have the effect of exempting from liability those employers who were historically the greatest offenders of the rights of blacks. A pattern or practice that would have constituted a violation of Title VII, but for the fact that the statute had not yet become effective, became a violation upon Title VII's effective date, and to the extent an employer continued to engage in that act or practice, it is liable under that statute. While recovery may not be permitted for pre-1972 acts of discrimination, to the extent that this discrimination was perpetuated after 1972, liability may be imposed.

Each week's paycheck that delivers less to a black than to a similarly situated white is a wrong actionable under Title VII, regardless of the fact that this pattern was begun prior to the effective date of Title VII.[39]

We now turn to the issue whether the Court of Appeals erred in upholding the District Court's refusal to accept the petitioners' expert statistical evidence as proof of discrimination [p. 398] by a preponderance of the evidence. In a case alleging that a defendant has engaged in a pattern and practice of discrimination under 707(a) of the Civil Rights Act of 1964,[40] plaintiffs

---

[38] Id. at 394.
[39] Id. at 395.
[40] 42 U.S.C. §2000e-6(a).

must "establish by a preponderance of the evidence that racial discrimination was the company's standard operating procedure — the regular rather than the unusual practice."[41]

At trial, petitioners relied heavily on multiple regression analyses designed to demonstrate that blacks were paid less than similarly situated whites. The U.S. expert prepared multiple regression analyses relating to salaries for the years 1974, 1975, and 1981. Certain of these regressions used four independent variables — race, education, tenure, and job title. Petitioners selected these variables based on discovery testimony by an Extension Service official that four factors were determinative of salary: education, tenure, job title, and job performance. In addition, regressions done by the Extension Service itself for 1971 included the variables race, sex, education, and experience; and another in 1974 used the variables race, education, and tenure to check for disparities between the salaries of blacks and whites. The regressions purported to demonstrate that in 1974 the average black employee earned $331 less per year than a white employee with the same job title, education, and tenure, and that in 1975 the disparity was $395. The regression for 1981 showed a smaller disparity which lacked statistical significance.[42]

The Court of Appeals stated: "[The] district court refused to accept plaintiffs' expert testimony as proof of discrimination by a preponderance of the evidence because the plaintiffs' expert had not included a number of variable factors the court considered relevant, among them being the across the board and percentage pay increases which varied from county to county. The district court was, of course, correct in this analysis."[43]

The Court of Appeals thought the District Court correct for two reasons: First, it rejected petitioners' regression analysis because it "contained salary figures which reflect the effect of pre-Act discrimination, a consideration not actionable under Title VII ...." Second, the court believed that "[a]n appropriate regression analysis of salary should ... include all measurable variables thought to have an effect on salary level." In particular, the court found that the failure to consider county-to-county

---

[41] Id. at 398 citing *Teamsters v. U.S.*, 431 U.S. 324, 336 (1977).
[42] 478 U.S. at page 399; citations and footnotes omitted from the opinion.
[43] 751 F.2d at 672.

differences in salary increases was significant. It concluded, noting: "[B]oth experts omitted from their respective analysis variables which ought to be reasonably viewed as determinants of salary. As a result, the regression analysis presented here must be considered unacceptable as evidence of discrimination." The Court of Appeals' treatment of the statistical evidence in this case was erroneous in important respects.

The Court of Appeals erred in stating that petitioners' regression analyses were "unacceptable as evidence of discrimination," because they did not include "all measurable variables thought to have an effect on salary level." The court's view of the evidentiary value of the regression analyses was plainly incorrect. While the omission of variables from a regression analysis may render the analysis less probative than it otherwise might be, it can hardly be said, absent some other infirmity, that an analysis which accounts for the major factors "must be considered unacceptable as evidence of discrimination." Normally, failure to include variables will affect the analysis' probativeness, not its admissibility.

Importantly, it is clear that a regression analysis that includes less than "all measurable variables" may serve to prove a plaintiff's case. A plaintiff in a Title VII suit need not prove discrimination with scientific certainty; rather, his or her burden is to prove discrimination by a preponderance of the evidence.[44] Whether, in fact, such a regression analysis does carry the plaintiffs' ultimate burden will depend in a given case on the factual context of each case in light of all the evidence presented by both the plaintiff and the defendant. However, as long as the court may fairly conclude, in light of all the evidence, that it is more likely than not that impermissible discrimination exists, the plaintiff is entitled to prevail.

In this case the Court of Appeals failed utterly to examine the regression analyses in light of all the evidence in the record. Looked at in its entirety, petitioners offered an impressive array of evidence to support their contention that the Extension Service engaged in a pattern or practice of discrimination with respect to salaries. In addition to their own regression analyses described above, petitioners offered regressions done by the

---

[44] *Texas Dept. of Community Affairs v. Burdine*, 450 U.S. 248, 252 (1981).

Extension Service for 1971 and 1974 that showed results similar to those revealed by petitioners' regressions. Petitioners also claim support from multiple regressions presented by respondents at trial for the year 1975. Using the same model that petitioners had used, and similar variables, respondents' expert obtained substantially the same result for 1975, a statistically significant racial effect of $384. Indeed, respondents also included in their analysis, "quartile rank" as an independent variable, and this increased the racial effect to $475.[45]

12.5.1 Collateral Evidence

Petitioners also presented evidence of pre-Act salary discrimination, and of respondents' ineffectual attempts to eradicate it. "In addition, the U.S. presented an exhibit based on 1973 data for 23 counties showing 29 black employees who were earning less than whites in the same county who had comparable or lower positions and tenure. Finally, and there was some overlap here with evidence used to discredit the county-to-county variation theory, petitioners presented evidence consisting of individual comparisons between salaries of blacks and whites similarly situated. Witness testimony, claimed by petitioners to be unrebutted, also confirmed the continued existence of such disparities.

Setting out the range of persuasive evidence offered by petitioners demonstrates the error of the Court of Appeals in focusing solely on the characteristics of the regression analysis. Although we think that consideration of the evidence makes a strong case for finding the District Court's conclusion clearly erroneous, we leave that task to the Court of Appeals on remand which must make such a determination based on the "entire evidence" in the record.[46]

12.5.2 Omitted Variables

To avoid summary judgment, those contesting a regression analysis must do more than list a few other variables that might have been included.

---

[45] Id. at 401. Citations omitted from the opinion.
[46] 478 U.S. 385, 404.

Instead, they must demonstrate that including an omitted variable would have eliminated any statistically significant disparity.

In *Smith v. VCU*, considered in Section 12.4.1, Michaell, Circuit Judge, dissented:[47]

> Trying to avoid summary judgment, plaintiffs (five male professors) simply threw rocks at VCU's salary equity study, rocks that either missed or glanced off harmlessly. On the record before us, VCU's multiple regression study establishes that there was a manifest imbalance in pay between men and women faculty members (women were underpaid) at that state university. Plaintiffs argue that VCU's study should have included more variables, but they fail to demonstrate that the inclusion of those variables would have eliminated the statistically significant effect of gender on salaries. Thus, plaintiffs' response does not satisfy either Supreme Court precedent (*Bazemore*) or settled summary judgment principles. With all respect for the majority, I therefore dissent because there is no material fact in dispute.

> VCU's study (a multiple regression analysis) mathematically estimated the effect of eight independent variables on salaries VCU paid to 770 tenured and tenure-track faculty members. The eight independent variables were: (1) national salary average (same discipline and rank), (2) doctorate or not, (3) tenure status, (4) quick tenure (within four years of appointment) or not, (5) years of experience at VCU, (6) academic experience before VCU, (7) service, if any, as department chair, and (8) gender. When the effect of the seven variables other than gender was taken into account, the effect of gender on faculty salaries was a statistically significant $1354 in 1989 and $1982 in 1991. Specifically, women faculty members, because they were women, on average were paid that much less than men. These findings were confirmed by VCU's expert statistician, Dr. Rebecca Klemm. Dr. Klemm testified (in deposition) that she took VCU's raw data and ran new regression studies using models different from the one used by VCU. Dr. Klemm's new studies consistently showed a gender difference in salary at VCU of a magnitude "very similar" to that found by VCU's original model. In moving for summary judgment, VCU relied on its original regression study and Dr. Klemm's testimony to establish the existence of a manifest imbalance in pay tied to

---

[47] 84 F.3d 672, 684 (4th Cir. 1996).

the impermissible factor of gender. Against this specific evidence, plaintiffs offered the deposition testimony and an affidavit (filed post-judgment) from Dr. Fred McChesney, an economics professor from Emory University. Unlike VCU's expert, Dr. McChesney did not perform a statistical study in this case. Indeed, he admitted that he had never performed a pay study such as the one at issue here. Nevertheless, Dr. McChesney opined that VCU's model should have included additional independent variables, such as performance and any prior service as an administrator. This omission was a fatal flaw, he said. When pressed, however, Dr. McChesney agreed that gender could affect salary at VCU. He also admitted that, even if the variables he suggested had been included, it was possible that the study results would have been essentially the same. In short, Dr. McChesney could not say that adding his variables would have had a statistically significant effect on the results of the VCU study. On this record the majority finds a genuine issue of material fact. In doing so, the majority gives Dr. McChesney far too much credit. Dr. McChesney's opinion does not justify the majority's conclusions that "failure to include a faculty member's status as a former administrator could easily have caused a salary differential that was not attributable to sex," and that "inclusion of the performance factors could very well alter the results of the multiple regression analysis." Dr. McChesney's untested conjecture does not undermine the validity of VCU's multiple regression analysis.

Bazemore is controlling here. In *Bazemore*, certain employees of the North Carolina Agricultural Extension Service sued various state and local officials, alleging that discriminatory differences in pay between black and white workers violated Title VII of the Civil Rights Act of 1964.[48] To help prove discrimination, the plaintiff-employees offered multiple regression analyses, which both the district court and our court rejected. The Supreme Court took the case and said that our court had erroneously concluded that "[a]n appropriate regression analysis of salary should ... include all measurable variables thought to have an effect on salary level."[49] After pointedly noting that the employees' expert testified that the disparities shown by the regressions were statistically significant, a unanimous Supreme Court said:

---

[48] 478 U.S. at 391.
[49] Id. at 399.

The [Fourth Circuit's] view of the evidentiary value of the regression analyses was plainly incorrect. While the omission of variables from a regression analysis may render the analysis less probative than it otherwise might be, it can hardly be said, absent some other infirmity, that an analysis which accounts for the major factors "must be considered unacceptable as evidence of discrimination."

Normally, failure to include variables will affect the analysis' probativeness, not its admissibility. Importantly, it is clear that a regression analysis that includes less than "all measurable variables" may serve to prove a plaintiff's case. A plaintiff in a Title VII suit need not prove discrimination with scientific certainty; rather his or her burden is to prove discrimination by a preponderance of the evidence.[50]

The majority cites this passage from *Bazemore* for the undisputed proposition that "a study must include all major factors." The majority then goes on to conclude that the variables omitted here (such as performance) are "crucial" factors because it believes their "inclusion ... could very well alter the results of [VCU's] multiple regression analysis." According to the majority, this creates a dispute of material fact "as to the validity of the study." The record does not support this conclusion. Plaintiffs never demonstrated the statistical significance of the omitted variables, and this is the fatal flaw in their response. It is fatal because *Bazemore* teaches that statistical significance must be the wedge that divides "major" (or "crucial") factors from other "measurable" factors. The *Bazemore* court criticized the governmental respondents for a trial strategy that made no attempt "— statistical or otherwise to demonstrate that when these [allegedly important] factors were properly organized and accounted for there was no significant disparity between the salaries of blacks and whites."[51] In addition, the Court noted three times that the regression analyses in the *Bazemore* record showed salary disparities that were "statistically significant."[52]

Bazemore therefore sends a clear signal: "major" factors are statistically significant factors.

---

[50] Citations and footnote omitted.
[51] 478 U.S. at 403, n. 14.
[52] Id. at 399, n. 9, 401, 404, n. 15.

In the case before us, the omitted variables cannot be characterized as "major" because plaintiffs did not demonstrate that their inclusion would eliminate the statistically significant disparity between the salaries of men and women at VCU. Thus, under *Bazemore* the district court was correct in rejecting plaintiffs' claim.

Even without *Bazemore*, VCU is entitled to summary judgment. Under settled Rule 56 jurisprudence, a defendant is entitled to judgment as a matter of law when the plaintiff fails to adduce facts on an element of his case for which he bears the burden of proof.[53] In this case, the majority acknowledges that plaintiffs bear the burden of establishing that VCU's affirmative action plan was implemented in the absence of a manifest imbalance in salary between male and female faculty members. To meet this burden, plaintiffs simply assert that VCU's multiple regression analysis failed to include enough factors. But, as I have already pointed out, plaintiffs have not shown the statistical significance of even one of the factors they say was erroneously excluded from VCU's study. In fact, their expert, Dr. McChesney, indicated that he did not know whether including the factors he identified would change the result of VCU's studies in a statistically significant way. To avoid summary judgment, plaintiffs must do more. The complaining male professors must produce firm evidence showing that, after adjusting for an omitted factor, a statistically significant gender effect on salary no longer exists. Nothing else can create a genuine issue of material fact concerning the study's validity. This case should be decided on the record. The record establishes that women faculty members at VCU were paid less because they were women. VCU made adjustments for the disparity based on a study that has not been discredited. The district court's award of summary judgment to VCU should be affirmed.

12.6 Alternate Forms of Regression Analysis

In this section, we consider several alternate forms of regression analysis including cohort analysis, ecological regression (the topic of Section 11.3.2), nonlinear and logistic regression, and the general problem of testing for significance.

---

[53] *Celotex Corp. v. Catrett*, 477 U.S. 317, 322 (1986); *Anderson v. Liberty Lobby, Inc.*, 477 U.S. 242, 248-249 (1986).

12.6.1 Cohort Analysis

In one form of *cohort analysis*, one extracts from the data a series of pairs, one male, one female (or one black, one white) *matched* on the basis of job classification, experience, education and any other factors of interest, and then compares their salaries using one of the tests described in Section 10.3. This approach was used in *Craig v. Minnesota State University Board*[54] but found irrelevant in that even if men and women of the same rank made comparable salaries, the failure to promote women in a timely fashion had created an existing inequality.

In another form, one divides the total sample into cohorts based on time and other relevant factors. Using a regression model provided by the plaintiffs, the district court in *Segar v. Civiletti*[55] found the Drug Enforcement Agency had discriminated against black agents in salary, promotion, initial grade assignments, work assignments, supervisory evaluation, and imposition of discipline.

In rebuttal, the agency provided a cohort analysis in which employees who started together at the same grade level in the same year were followed as a group. The agency broke the sample up into 15 separate groups or cohorts and found no significant discrimination in any of these groups after some individuals were reclassified. The court concluded this was because the resultant groups were too small to generate statistically significant evidence of discrimination and ruled against the agency.[56]

12.6.2 Linear, Nonlinear, and Logistic Regression

A formula such as $Y = b_0 + b_1 X^2$ is known to statisticians as *linear regression*, even though the curve it represents is far from a straight line. The coefficients the computer is to estimate, b_0 and b_1, are in linear form. In economics and pharmacokinetics, one often encounters equations of the form $Y = b_0 + b_1 f[X, b_2]$ where the unknown coefficient b_2 enters into the formula in a nonlinear and implicit fashion. While many methods exist for the solution of such *nonlinear regression* equations,[57] they do not all yield the same solution. Worse, they may yield different values of the coefficients, even with the same method, if the test starts out with different initial "guesstimates" of the unknown parameters. With so many potential objections, nonlinear regression is not quite ready for courtroom use.

---

[54] 731 F.2d 465 (8th Cir. 1984).
[55] 508 F. Supp. 690. (D.D.C. 1981); *Segar v. Smith* 738 F.2d 1249 (D.C. Cir. 1984), <u>cert denied</u>, 105 S.Ct. 2357 (1985).
[56] Ibid. at 712.
[57] See Gallant [1987] for example.

Logistic regression, by contrast, is a well-defined method for fitting and predicting the values of a variable *Y* which takes only two possible values, the value 1, representing success or survival, with probability *p*, and the value 0, representing failure or death, with probability $1 - p$; *p* is the dependent variable. The corresponding logistic regression equation takes the form:

$$\log [p/(1 - p)] = b_0 + b_1 f_1[X_1] + b_2 f_2[X_2] + \ldots + b_n f_n[X_n]$$

and estimates of the regression parameters may be obtained by the methods of ordinary least squares.

12.6.3 Testing for Significance

Suppose we wish to determine whether there exists a significant difference in the employment (or promotion or salary) of individuals over 40 years of age and individuals who are younger (or differences between men and women or Hispanics and non-Hispanics). We might do any of the following:

1. Develop a multiple regression function using an indicator (0,1) variable to represent the age group (sex, race). After correcting for the effects of the other variables, test whether or not the age (sex, race) effect is significant.
2. Develop two separate multiple regression functions, one for each age group (sex, race). Compare the predicted employment rates (promotion rates, salaries).
3. Develop a single multiple regression function for both age groups. Compare the residuals for the under-40 individuals with the residuals for the over-40 group using a permutation test.

All three methods rely on two assumptions concerning the error terms: (1) they are independent and (2) they have the same distribution.

The first approach is the most commonly used. Its drawbacks lie in the use of an indicator variable (so that the calculated significance levels are only approximations) and its reliance on an assumption of normality for the error terms. Although the plaintiffs in *Wilkins v. University of Houston*[58] were able to demonstrate a statistically significant pay differential between the sexes of $694, the court rejected their argument, siding with the university. "The multiple regression analyses do not indicate

---

[58] 654 F.2d 388 (5th Cir. 1981), rehearing denied, 662 F.2d 1156 (1981), vacated, 459 U.S. 809 (1982), on remand, 695 F.2d 134 (1983).

discrimination against women because the model with sex as a factor explained only 0.8% more of the total variation around the average salary than did the model without sex."[59] R^2 = 52.4% without sex as a factor and 53.2% considering sex as a factor.

The second approach has the advantage that the court may compare the coefficients of the various explanatory factors between the two groups and thus pinpoint the sources of any discrimination.

The third approach, which we favor because it requires the fewest assumptions and produces exact significance levels, was accepted by the district court in *Sobel v. Yeshiva University*[60] but rejected by the Second Circuit Court of Appeals.[61] This approach was accepted by the administrative law judge in *U.S. Department of Labor v. Harris Trust and Savings Bank.*[62]

12.7   When Statistics Don't Count

12.7.1   Age Discrimination

Little remains today of the Age Discrimination in Employment Act[63] as the Supreme Court has chosen to construe it so narrowly as to make it all but unenforceable.

Age as a Continuous Variable

In *Kroger v. Reno*,[64] a class of older Deputy U.S. Marshals alleged age discrimination by the FBI in violation of the Age Discrimination in Employment Act. Among the items offered into evidence was a set of correlation results demonstrating that on certain components of the promotion package, such as training, education, FIT assessment, and annual appraisal, the older an applicant the more likely his scores were to be low.

> The sole evidence offered in support of discrimination at the stage of ultimate promotion was plaintiffs' regression analysis purporting to show that, holding aggregate scores constant, the younger deputies had a greater chance of promotion. The district court rejected this on the ground that the regression was based

---

[59] Id. 403-404 and n. 19.
[60] 566 F. Supp. 1166, 1169-1170 (S.D. N.Y. 1983); Id. at 1183, n. 42.
[61] 839 F.2d 18, 35-36 (1983).
[62] No. 78-OFCCP-2 (ALJ decision, Dec. 22, 1986).
[63] 29, U.S.C. §623, 633a et seq.
[64] 98 F.3d 631 (D.C. Cir. 1996).

simply on variation in age, as opposed to discrimination against those 40 or over. Because the expert used a continuous variable for age, her results do not address the issue of whether, holding scores constant, a deputy 40 years old or more is less likely to be promoted than a deputy under 40. All the regression shows is that older deputies (of any age) are less likely to be promoted, relative to younger deputies (of any age). But the entire statistical advantage of the younger deputies could have come from disparate promotion rates as between deputies in the under-40 category. Because these deputies are not protected under the ADEA, regardless of demonstrated discrimination, the inclusion of this data is fatal to the expert's conclusion. See *Murnane v. American Airlines, Inc.*,[65] (employer's guideline against hiring persons over 30 considered only insofar as it was applicable to those over 40, because those under 40 are not protected under the ADEA); see also *Paetzold and Willborn*[66] (observing that it would be error to infer illegal discrimination from figures simply showing statistical significance in "the relationship between age and termination for all employees").

Plaintiffs urge that despite this defect, the court should have accepted the regression as having some probative value. They point especially to *Bazemore v. Friday*,[67] in which the Court held that the lower courts erred in rejecting the plaintiffs' regression analysis. The analysis had demonstrated a wage disparity between black and white employees with the same job title, education and tenure. The district court had rejected the regression because there were other variables, such as county-by-county wage variations, that might have accounted for the salary disparity. The Supreme Court rejected this argument, holding that "the omission of variables from a regression analysis may render the analysis less probative than it otherwise might be," but that it does not make the analysis unacceptable as evidence.[68] The Court also said that the defendants had "made no attempt ... to demonstrate that when these factors were properly organized and accounted for there was no significant disparity between the salaries of blacks and whites."[69]

---

[65] 667 F.2d 98, 99-100 and n. 3 (D.C. Cir. 1981).

[66] §7.07 at 7-12 (1995).

[67] 478 U.S. 385 (1986).

[68] Id. at 400.

[69] Id. at 403-404 n. 14.

Following *Bazemore*, courts have taken the view that a defendant cannot undermine a regression analysis simply by pointing to variables not taken into account that might conceivably have pulled the analysis's sting.[70] Courts have not, however, understood *Bazemore* to require acceptance of regressions from which clearly major variables have been omitted, such as education and prior work experience,[71] or, in decisions on academic pay, rank and tenure, the quality of teaching and research, and community and institutional service.[72]

Here, however, we do not deal with a regression that simply omits a variable of potential significance. Instead we have one that (if valid at all) supports an inference that is not legally relevant: that, holding aggregate scores constant, variations in age over the entire age range of applicants, are statistically associated with promotion. It thus fails to show a disparity that disfavors deputies 40 or older. To have required the defendant to have constructed and conducted the proper analysis to correct the plaintiffs' error would be to improperly shift the burden of proof.

Nor does the other aspect of *Bazemore's* analysis of statistical proof suggest that the court should have given plaintiffs' regression any weight. The decision considered whether the inclusion of pre-Title VII data invalidated the plaintiffs' statistical analysis and concluded that it did not, because "proof that an employer engaged in racial discrimination prior to the effective date of Title VII might in some circumstances support the inference that such discrimination continued, particularly where relevant aspects of the decision-making process had undergone little change."[73]

The inclusion of pre-Title VII data might be thought analogous to the use of the continuous variable for age in our case, in the sense that both involve the inclusion of data that relate to unactionable discrimination. In *Bazemore*, however, the inclusion

---

[70] See, e.g., *Palmer v. Shultz*, 815 F.2d 84, 106 (D.C. Cir. 1987) (possible impact of individual preferences insufficient to justify rejection of plaintiffs' analysis); *Segar v. Smith*, 738 F.2d 1249, 1277 (D.C. Cir. 1984) (similar); *EEOC v. General Telephone Co.*, 885 F.2d 575, 582 (9th Cir. 1989) (similar); *Sobel v. Yeshiva Univ.*, 839 F.2d 18, 33-34 (2nd Cir. 1988) (similar).

[71] *Sheehan v. Purolator, Inc.*, 839 F.2d 99, 103 (2nd Cir. 1988).

[72] *Penk v. Oregon State Bd. of Higher Educ.*, 816 F.2d 458, 464-465 (9th Cir. 1987) (distinguished in *General Telephone*, 885 F.2d at 581-582).

[73] Id. at 402; see also *Valentino v. U.S. Postal Service*, 674 F.2d 56, 71, n. 26 (D.C. Cir. 1982).

of pre-Title VII data provided information about an employer's treatment of the protected class (just at a different time), whereas statistical disparities within the under-40 category say nothing about treatment of the protected class."[74]

A Welcome Exception

The Sixth Circuit is willing to consider evidence of age-related discrimination apart from the division at age 40; see *Barnes v. GenCorp, Inc.*[75]

Hazen Paper: Absent a Smoking Gun

The management of Hazen Paper determined to trim its costs by eliminating employees who would shortly be claiming their pensions if allowed to continue with the company. Not unreasonably, the terminated employees felt they were discriminated against on the basis of age. The court thought otherwise, ruling it insufficient for an age discrimination plaintiff to offer proof that some factor correlated with age, but analytically distinct, had a casual impact on an employer's decision to terminate, absent proof that the employee's age actually influenced the decision.[76]

12.7.2 Gender Discrimination

Oklahoma passed a statute in 1958 that would have let women between 18 and 21 drink 3.2% beer, but not allow men in the same age group to do so. An indignant male sued the state of Oklahoma. The state collected statistics that showed that 2% of the males and fewer than 0.2% of the females in that age group had been arrested for alcohol-related driving offenses. That means 427 out of 69,688 males and only 24 out of 68,500 females were arrested. Justice Brennan was unimpressed by the disparity and wrote in *Craig v. Born*,[77] "While such a disparity is not trivial in a statistical sense, it hardly can form the basis for employment of a gender line as a classifying device."

12.7.3 Sentencing

David Baldus' landmark study of sentencing in the state of Georgia was based on over 2000 murder cases in that state during the 1970s.[78] He and

[74] 98 F.3d 631, 638.
[75] 896 F.2d 1457 (6th Cir.), 498 U.S. 878 (1990).
[76] *Hazen Paper Co. v. Biggins*, 507 U.S. 604 (1993), 123 L.Ed.2d 338, 113 S.Ct. 1701.
[77] 429 U.S. 190 (1976) 201-202.
[78] See discussion in Section 12.4.

his colleagues, Charles Pulaski and George Woodworth, subjected his data to an extensive analysis, taking account of 230 variables that could have explained the disparities on nonracial grounds. One of his models concludes that, even after taking account of 39 nonracial variables, defendants charged with killing white victims were 4.3 times as likely to receive death sentences as defendants charged with killing blacks. Regardless, the court felt:

> In light of the safeguards designed to minimize racial bias in the process, the fundamental value of jury trial in our criminal justice system, and the benefits that discretion provides to criminal defendants, we hold that the Baldus study does not demonstrate a constitutionally significant risk of racial bias affecting the Georgia capital sentencing process.[79]

> Our analysis begins with the basic principle that a defendant who alleges an equal protection violation has the burden of proving "the existence of purposeful discrimination."[80] Statistics at most may show only a likelihood that a particular factor entered into some decisions. There is, of course, some risk of racial prejudice influencing a jury's decision in a criminal case. There are similar risks that other kinds of prejudice will influence other criminal trials. The question is "at what point that risk becomes constitutionally unacceptable."[81,82]

> McCleskey asks us to accept the likelihood allegedly shown by the Baldus study as the constitutional measure of an unacceptable risk of racial prejudice influencing capital sentencing decisions. This we decline to do.

> It is the jury that is a criminal defendant's fundamental "protection of life and liberty against race or color prejudice."[83] Specifically, a capital sentencing jury representative of a criminal defendant's community assures a "diffused impartiality,"[84] in the jury's task of "express[ing] the conscience of the community on the ultimate question of life or death."[85] Individual jurors bring to their deliberations "qualities of human nature and varieties of human

---

[79] *McCleskey v. Kemp*, 481 U.S. 279, 314 (1987).

[80] *Whitus v. Georgia*, 385 U.S. 545, 550 (1967).

[81] Turner v. Murray, 476 U.S. 28, 36, n. 8 (1986).

[82] 481 U.S. 279, 309.

[83] *Strauder v. West Virginia*, 100 U.S. 303, 309 (1880).

[84] *Taylor v. Louisiana*, 419 U.S. 522, 530 (1975) (quoting *Thiel v. Southern Pacific Co.*, 328 U.S. 217, 227 (1946) (Frankfurter, J., dissenting)).

[85] *Witherspoon v. Illinois*, 391 U.S. 510, 519 (1968).

experience, the range of which is unknown and perhaps unknowable."[86] The capital sentencing decision requires the individual jurors to focus their collective judgment on the unique characteristics of a particular criminal defendant. It is not surprising that such collective judgments often are difficult to explain. But the inherent lack of predictability of jury decisions does not justify their condemnation. On the contrary, it is the jury's function to make the difficult and uniquely human judgments that defy codification and that "buil[d] discretion, equity, and flexibility into a legal system."[87] McCleskey's argument that the Constitution condemns the discretion allowed decision makers in the Georgia capital sentencing system is antithetical to the fundamental role of discretion in our criminal justice system. Discretion in the criminal justice system offers substantial benefits to the criminal defendant. Not only can a jury decline to impose the death sentence, it can decline to convict or choose to convict of a lesser offense. Whereas decisions against a defendant's interest may be reversed by the trial judge or on appeal, these discretionary exercises of leniency are final and unreviewable. Similarly, the capacity of prosecutorial discretion to provide individualized justice is "firmly entrenched in American law."[88] As we have noted, a prosecutor can decline to charge, offer a plea bargain, or decline to seek a death sentence in any particular case.[89] Of course, "the power to be lenient [also] is the power to discriminate,"[90] but a capital punishment system that did not allow for discretionary acts of leniency "would be totally alien to our notions of criminal justice."[91]

At most, the Baldus study indicates a discrepancy that appears to correlate with race. Apparent disparities in sentencing are an inevitable part of our criminal justice system. The discrepancy indicated by the Baldus study is "a far cry from the major systemic defects identified in Furman."[92,93]

A similar finding was made by a Georgia court in *Stephens v. State.*[94]

---

[86] *Peters v. Kiff*, 407 U.S. 493, 503 (1972) (opinion of Marshall, J.).
[87] Kalven and Zeisel [1966].
[88] LaFave and Israel [1984].
[89] Footnotes omitted.
[90] Davis [1973] p. 170.
[91] *Gregg v. Georgia*, 428 U.S. 200, n. 50.
[92] *Pulley v. Harris*, 465 U.S. 37 (1984).
[93] 481 U.S. 279, 313; footnotes omitted.
[94] 456 S.E.2d 560 (Ga. 1995).

12.8   Summary

The courts distrust arguments predicated on the presence or absence of a single factor. Ideally, a regression model will include all relevant factors. Nonetheless, parties challenging a model cannot undermine a regression analysis simply by pointing to variables not taken into account that might conceivably have pulled the analysis's sting. Rather, that party must clearly identify major variables, the omission of which renders the analysis suspect.

Failure to include variables will affect an analysis' probativeness, not its admissibility. Collateral evidence in conjunction with statistical evidence can be persuasive.

It is only when a model bears no rational relationship to the characteristics of the data to which it is applied that courts will hold the use of the model was arbitrary and capricious.

Courts are reluctant to rely solely on a statistic such as R (defined in Section 11.1) to choose one model over another. Although, the effect on R^2 of the addition or deletion of a factor provides a measure of a factor's relevance, this effect will depend upon the numbers and kinds of other variables and the order in which they are added to or deleted from the model.

Chance is an inevitable part of real world observation. A regression line does not explain a phenomenon completely, but does so only in a general way. While each data point will be associated with some residual (the difference between actual and fitted values), so long as this residual is within acceptable statistical limits, the fact that some data points necessarily fall above or below the line does not render the regression analysis invalid.

Points of attack on a regression analysis include:

■ How the data was collected and maintained; its quality and quantity
■ The forms of the equations and the variables that go into them
■ The methodology used to determine the coefficients of the equation
■ The application of the resulting equations
■ The lack of evidence of validation of model

When comparing several groups, the following procedure is recommended. Develop a single multiple regression function for all groups. Compare the residuals for one group with the residuals for the others using a permutation test. This, like any other statistical procedure may be subject to attack on the ground that the appropriate population was not used for comparison purposes.

APPLYING STATISTICS IN THE COURTROOM

In this final part, we review the application of statistics in the courtroom. The emphasis of Chapter 13 is on preventive measures and counterattacks. Chapter 14 describes the trial process for the benefit of the statistician who may encounter it in its entirety for the first time. For the benefit of the attorney, Chapter 15 describes how to get the most effective use from statistics and the statistician throughout a trial and includes a list of questions to use during discovery.

Chapter 13

Preventive Statistics

13.1 Concepts

How many times have you thought (if not actually voiced the thought), "If only they'd consulted me *before* they bought that property, signed that contract, published that article, distributed that chemical." In this chapter, we consider the many statistically based preventive actions your clients might take along with the many defensive motions that might be made once it is recognized that your client is a victim of coincidence or bad statistics.

Concepts introduced include *controls*, *power of a statistical test*, *coincidence*, and *small-sample variability*. We also reemphasize the importance of using random, representative samples.

13.2 Appropriate Controls

A basic principle of experimental design is to utilize control subjects. In establishing evidence of discriminatory intent or providing baseline data to be used in the award of damages, we also need to refer to control or reference populations. The court's views on controls are examined in the next several sections.

13.2.1 Breast Implants

Who knows (or will admit) what executive or executive committee at Dow Corning first decided it was not necessary to do experimental studies on silicone implants because such studies were not mandated by government regulations? It is terrifying to realize that the first epidemiological study of whether breast implants actually increase the risks of certain diseases and symptoms was not submitted for publication until 1994, whereas the first modern implants (Dow Corning's Silastic mammary prosthesis) were placed in 1962.[1] The first successful lawsuit in 1984 resulted in a jury award of $2 million. Award after award followed with the largest ever, more than $7 million, going to a woman whose symptoms had begun even before she received the implants.[2]

Every study involves at least two groups of subjects — those who took the drug or implanted the device and those who did not, the so-called "control group." When I started work at the Upjohn Company several decades ago, I asked my boss how many controls he thought should be used. The answer was a surprise. He recommended the use of twice as many subjects in each control group as the number devoted to experimental treatment, and that two types of controls be used, positive (e.g., aspirin v. Motrin®) and neutral (e.g., placebo v. Motrin).

Unlike the Dow Corning executives, he was concerned with both the long- and short-term costs. The reason for using so many subjects in each control group was that, "Life is full of surprises: you leave work one day whistling, the next day you're back with a head cold. Most of the time these negative effects have nothing to do with the treatment. By using many control subjects, you ensure the normal wear and tear of ordinary life will be detected and accounted for and won't be falsely associated with the treatment you're trying to investigate."

The controversy over silicon implants provides definitive proof of my ex-boss' wisdom. Major corporations went bankrupt. In late 1998, Dow Corning Co, agreed to pay $3.2 billion to settle claims from more than 170,000 women even though, less than two weeks later, a panel of respected and neutral scientists concluded there was no credible evidence that silicone gel implants cause disease.[3] All this was because a basic principle of experimental design had been ignored.

A jury in Texas awarded $25 million to Pamela Johnson although she had no symptoms of the type associated with silicone gels. In the *Hopkins* case, with more than $7 million at stake ($840,000 in compensatory damages and $6.5 million in punitive damages), Dow Corning argued that the district court erred in admitting testimony that was not based on scientifically acceptable principles (i.e., case control or cohort studies). The 9th Circuit appeals court responded that, "The record reflects that Hopkins' experts based their opinions on the types of scientific data and

---

[1] According to Marcia Angell [1996], the recipient still has her original implants and has no complaints.

[2] *Hopkins v. Dow Corning Corp.*, 33 F.3d 1116 (9th Cir. 1994).

[3] *Los Angeles Times*, Dec. 3, 1998, p. B10.

utilized the types of scientific techniques relied upon by medical experts in making determination regarding toxic causation *where there is no solid body of evidence to review.*"[4]

Do not allow your clients to become so vulnerable. Recommend they use controls in their studies.

13.2.2   A Basis for Comparison

Section 1.2 discussed the importance of finding the appropriate population for comparison purposes. The next two cases further illustrate the courts' view of that issue.

The owner of the Red Turtle Bar complained that her business was subjected to unfair and discriminatory treatment by Santa Ana police officers and claimed she had been denied equal protection under the law.[5]

"The first step in equal protection analysis is to identify the [defendants'] classification of groups."[6] To accomplish this, a plaintiff can show that the law is applied in a discriminatory manner or imposes different burdens on different classes of people.[7]

Once the plaintiff establishes governmental classification of groups, it is necessary to identify a "similarly situated" class or control group against which the plaintiff's class can be compared. "Discrimination cannot exist in a vacuum; it can be found only in the unequal treatment of people in similar circumstances."[8] "The goal of identifying a similarly situated class … is to isolate the factor allegedly subject to impermissible discrimination. The similarly situated group is the control group."[9]

Freeman, the owner of the Red Turtle, defined her class as "Mexican immigrant bars" with crime problems, regardless of license type. She attempted to define the similarly situated control group as "non-Mexican immigrant bars" with crime problems. The district court determined that for purposes of defining the similarly situated class, Freeman could only introduce evidence of premises that had the same license type as the Red Turtle. Freeman argued that the district court's classification focused on an irrelevant similarity between license types, while ignoring the relevant factors of the patrons' races or national origins.

In her offer of proof, Freeman indicated that 17 of 18 bars targeted for closure by the Santa Ana Police Department were Mexican immigrant

---

[4] Ibid. at 1124. Italics are mine.

[5] *Freeman v. City of Santa Ana*, 68 F.3d 1180 (9th Cir. 1995).

[6] *Country Classic Dairies, Inc. v. State of Montana*, 847 F.2d 593, 596 (9th Cir. 1988).

[7] *Christy v. Hodel*, 857 F.2d 1324, 1331 (9th Cir. 1988), <u>cert. denied</u>, 490 U.S. 1114 (1989).

[8] *Attorney General v. Irish People, Inc.*, 684 F.2d 928, 946 (D.C. Cir. 1982), <u>cert. denied</u>, 459 U.S. 1172 (1983).

[9] *U.S. v. Aguilar*, 883 F.2d 662, 706 (9th Cir. 1989), <u>cert. denied</u>, 498 U.S. 1046 (1991).

bars, only two of which held licenses of the same type as hers. She based her selective prosecution claim on the evidence that there were 8 to 12 non-Mexican immigrant bars with similar crime problems that the department did not attempt to close.

Noting that a district court has broad discretion to exclude evidence,[10] her appeal was rejected.

13.2.3 Extent of Damages

In determining the extent of damages, the members of the control group should be similar to the injured party in all aspects but one, the one being the damage allegedly caused by the defendant. Such factors might include age, race, sex, or, as in *Penney v. Praxair*,[11] age and time since withdrawal from medication.

Leonard Penney was sleeping in the front passenger seat of a car that was rear-ended by a loaded tanker truck owned by Praxair, Inc. After both MRI and CT scans detected no brain injury, a positron emission tomography (PET) scan was made of his brain. To detect abnormalities, the PET scan of his brain needed to be compared with the PET scans from a control group.[12] The control group consisted of 31 persons, with ages ranging from 18 to 70. Penney's PET scan showed brain abnormalities consistent with a traumatic brain injury and plaintiffs intended to use this testimony to prove the existence of a closed head injury.

Praxair filed a motion *in limine* to exclude the PET scan evidence, and argued it was not reliable enough to withstand analysis under the Supreme Court's decision in *Daubert*.[13] The district court excluded the PET scan results, reasoning that the evidence would not be helpful to the jury in deciding the issues when compared with the likelihood that the jury would misapply the evidence. Upholding this decision, the Eighth Circuit stated:

> General acceptance in the scientific community is no longer a precondition to the admission of scientific evidence.[14] However, a trial judge must still ensure that "an expert's testimony both rests on a reliable foundation and is relevant to the task at hand."[15] "This entails a preliminary assessment of whether the reasoning or methodology underlying the testimony is scientifically valid

---

[10] *LuMetta v. U.S. Robotics, Inc.*, 824 F.2d 768, 770 (9th Cir. 1987).

[11] 116 F.3d 330 (8th Cir. 1997).

[12] *Hose v. Chicago Northwestern Transp. Co.*, 70 F.3d 968, 973 (8th Cir. 1995).

[13] *Daubert v. Merrell Dow Pharm., Inc.*, 509 U.S. 579 (1993).

[14] Ibid. at 597.

[15] Id.

and of whether that reasoning or methodology properly can be applied to the facts in issue."[16] In this case, plaintiffs failed to establish a sufficient foundation to support the admission of the PET scan evidence.

According to the parties' submissions, PET scan results can be affected by a person's age, medical history and medications. Because Leonard was sixty-six years old at the time of the scan, it is not clear from the record exactly how accurate a comparison this control group could provide. Furthermore, although persons are normally instructed to remain off medication for seven days prior to the administering of a PET scan, Leonard submitted to the test while still taking his regular medications for his heart condition and other maladies. None of the other control-group subjects was on medication at the time of their PET scans. It is not clear whether these factors had any effect on the test results. However, it was plaintiffs' burden to establish a reliable foundation for the PET scan readings. On these facts, plaintiffs did not make such a demonstration and it was within the district court's discretion to exclude the evidence.

As the plaintiffs point out, we have previously upheld the admission of PET scan evidence.[17] However, because the admission of scientific evidence in one case does not automatically render that evidence admissible in another case, we assume that *Hose* did not present the same evidentiary problems as does this case.

13.2.4 Placebo Effect

Good experimental design requires the use of representative random samples and negative (placebo) controls.

A seller may not represent a product as "effective" when its efficacy results solely from a "placebo effect."[18] The representation of effectiveness

---

[16] Id. at 592-593.

[17] See *Hose*, 70 F.3d at 973.

[18] The term *placebo effect* refers to the fact that even a product of no inherent merit whatsoever will often have some degree of effectiveness in treating the condition for which it is employed, for psychological or other reasons. For example, a patient who ingests sugar pills while believing that they are strong pain relievers may well experience some pain relief, even though sugar pills are inherently worthless in treating pain. In this example, the sugar pill is a placebo and the relief experienced by the patient is the placebo effect. *F.T.C. v. Pantron I Corp.*, 33 F.3d 1088, n. 1 (9th Cir. 1994).

constitutes a false advertisement under the Federal Trade Commission Act. Pantron I Corporation marketed a product known as the Helsinki Formula. This product supposedly arrested hair loss and stimulated hair regrowth in baldness sufferers; it consisted of a conditioner and a shampoo, and sold at a list price of $49.95 for a 3-month supply. The ingredients that allegedly caused the advertised effects were polysorbate 60 and polysorbate 80. Pantron offered a full money-back guarantee to those not satisfied with the product. Its late-night infomercials featured both the hair loss claim and the claim that the formula promoted growth of new hair in baldness sufferers. It also represented that recognized scientific studies supported these claims.

In response, the U.S. Postal Service, the Food and Drug Administration, the Los Angeles County District Attorney, and even the Council of Better Business Bureaus took varying degrees of action against Pantron's advertising and marketing of the formula. The FTC's complaint was directed to the advertisements that represented that the Helsinki Formula was effective and that Pantron had scientific support for this conclusion. The complaint alleged that the representations were false and constituted an unfair or deceptive trade practice in violation of Sections 5 and 12 of the Federal Trade Commission Act.[19] The FTC sought a permanent injunction and monetary equitable relief.

At trial, the FTC presented a variety of evidence that tended to show that the Helsinki Formula had no effectiveness (other than its placebo effect) in arresting hair loss or promoting hair regrowth. It introduced the expert testimony of a dermatologist who stated that, based on his knowledge and review of the medical literature, there was "no reason to believe" that the Helsinki Formula would be in any way useful in treating hair loss. He also stated that his opinion was in accord with the consensus view of the medical community.

A second expert stated that the studies on which Pantron relied failed to satisfy the generally accepted scientific standards of being *randomized, double-blinded,* and *placebo controlled.*

We learned about the need for randomization in Chapter 2. In *placebo-controlled* studies, we have both a treatment group and a control group, the latter receiving a treatment that looks, feels, and, perhaps, tastes like the real thing but is actually a biologically inert substance or filler. A *double-blind* study is one in which neither the patient nor the experimenter knows which treatment the patient is receiving. The rationale for keeping the physician unaware is that if he knew the pill he was administering was only a harmless filler (a placebo), he might tend to be careless in its administration and communicate his indifference to the patient. We know

---

[19] See 15 U.S.C. 45, 52.

that the mere suggestion to "take this pill and you'll feel better in a few days" often works wonders. This placebo effect would not occur if we knew or even suspected we were in the control group.

A third FTC expert, who had conducted a study of another polysorbate 60-based baldness treatment, expressed his opinion that neither polysorbate 60 nor polysorbate 80 — the two allegedly result-producing ingredients in the Helsinki Formula — was effective in reducing hair loss or promoting hair regrowth.

The court also took judicial notice that the Food and Drug Administration issued a rule that prohibited marketers of over-the-counter baldness treatments from labeling their products as effective.[20] The FDA's final rule, which applies to all over-the-counter hair growth products, specifically identifies polysorbate 60 and several other ingredients found in the Helsinki Formula.[21] The FDA rule concludes that "[b]ased on evidence currently available, all labeling claims for OTC hair grower and hair loss prevention drug products for external use are either false, misleading, or unsupported by scientific data."

The FTC and its experts introduced evidence of two studies that determined that polysorbate-based products were ineffective in stopping hair loss and promoting hair regrowth. The more important study was a placebo-controlled, double-blinded, randomized study published in the *Archives of Internal Medicine*, a peer-reviewed journal.

This study found "[n]o statistically significant difference" between the control and treatment groups; nearly a quarter of the participants in each group reported new hair growth. The authors concluded that "polysorbate 60 is not an effective remedy for MPB [male pattern baldness]," and that hair regrowth products possess a very strong placebo effect.[22]

In addition, the FTC introduced the so-called Shuster study, an unpublished study that compared a polysorbate-based product to Pantene, a hair product presumed to have no inherent curative or restorative qualities. This study also concluded that polysorbate-based products were ineffective, although the FTC acknowledges that "the failure to include a clear-cut placebo somewhat reduces [its] value."[23]

In response, Pantron introduced evidence that users of the Helsinki Formula were satisfied that it was effective. It offered the live and deposition testimony of 18 users who experienced hair regrowth or a reduction in hair loss after using the formula. It also introduced evidence of a "consumer satisfaction survey" it conducted in late 1988. In this survey,

---

[20] See 21 C.F.R. 310.527.

[21] Id. 310.527(a).

[22] A possible Pantron defense was that the sample in the referenced study was not large enough to detect the effect. See Section 13.3.

[23] *FTC v. Pantron I Corp.*, 33 F. 3d 1088, 1093 (9th Cir. 1994).

which was taken during routine sales follow-up calls, a representative of Pantron interviewed a cross-section of 579 Helsinki Formula customers. Although the Pantron official who conducted the survey could not remember the questions he asked, and the company did not keep a record of these questions, Pantron introduced the results into evidence. The survey data showed positive results in a significant percentage of users, ranging from 29.4% of those who had used the product less than 2 months, to 70% of those who had used it for 6 months or more. Pantron also introduced evidence that over half of its orders came from repeat purchasers, that it had received very few written complaints, and that very few customers (fewer than 3%) exercised their rights under the money-back guarantee.[24]

Pantron also introduced several clinical studies of its own. First, it offered the results of Finnish studies (for which the Helsinki Formula was named) performed by Dr. Ilona Schreck-Purola. Her uncontrolled, unblinded, unrandomized, un-peer-reviewed study concluded that a polysorbate-based product was effective in arresting excessive hair loss within 2 to 4 weeks, and that it led to new hair growth in 60% of the subjects within 4 months. Although Dr. Schreck-Purola acknowledged that "the medical community remains of the opinion that polysorbates are not effective in treating male pattern baldness," she nonetheless stated that, in her opinion, "polysorbates help alleviate baldness by destroying the cholesterol in the testosterone that destroys hair follicles."

District Court Findings

The district court found that Pantron had made the representations of efficacy and scientific support that the FTC had alleged, but determined that "[t]here is no evidence in the record to support a contention that the Helsinki Formula is wholly ineffective."[25] The district court found that the studies and anecdotal evidence offered by Pantron "support[ed] the proposition that the compound works for some people some of the time."[26] Thus, it concluded that the FTC had failed to carry its burden of showing that Pantron made a false claim when it represented that the Helsinki Formula was effective.

However, the district court found "no scientifically valid evidence that polysorbate 60 is effective for treatment of hair loss or for inducing growth."[27] Thus, the district court concluded that the FTC had "marginally

---

[24] In John Grisham's fictional *Runaway Jury*, we learn you can always find an expert available for hire. Maybe, but there is no guarantee the court will find the evidence compelling or be willing to let the jury listen to it.

[25] *FTC v. Pantron I Corp.*, 33 F. 3d 1088, 1094 (9th Cir. 1994).

[26] Ibid.

[27] Ibid.

carried its burden on the charge of falsity in defendant's claims of scientific proof." Accordingly, it entered an injunction, that barred Pantron from making any express or explicit representations that scientific evidence establishes that the Helsinki Formula "is effective in any way in the treatment of baldness or hair loss." The order specifically allowed the defendants to state that the Helsinki Formula (or a similar product) was the subject of medical investigative work by responsible European physicians, if such statement was accompanied by a clear and conspicuous disclosure that the work did not conform to recognized standards in the U.S. for medical and scientific studies.

Another provision of the injunction prohibited "any misrepresentation … regarding the effectiveness of such product or program in the treatment of baldness," but it allowed Pantron to state that the Helsinki Formula is effective to some extent for some people in dealing with male pattern baldness, if such statement was accompanied by a clear and conspicuous disclosure that the product's effectiveness (1) is more likely to involve arrest of hair loss than growth of new hair, and (2) is not explained or supported by scientific studies recognized under standards in use in the U.S.

The court refused to order monetary equitable relief, because it concluded "[t]he FTC has not established that defendant's conduct caused actual deception and injury to consumers, nor that the defendant knew or should have known such conduct was fraudulent."[28]

Appeals Court Ruling

The appeals court held that the district court erred in concluding that Pantron's representations regarding the Helsinki Formula's efficacy did not amount to false advertising.

> Although there was sufficient evidence in the record to support the district court's finding that use of the Helsinki Formula might arrest hair loss in some of the people some of the time, the overwhelming weight of the proof at trial made clear that any effectiveness is due solely to the product's placebo effect. As we explain *infra*, we conclude that a claim of product effectiveness is "false" for purposes of section 12 of the Federal Trade Commission Act if evidence developed under accepted standards of scientific research demonstrates that the product has no force beyond its placebo effect.

---

[28] Ibid.

As [the FTC's three experts] Drs. Kramer, Orenberg, and Ganiats testified, "the consensus of the medical and scientific community is that polysorbate-based products have no effectiveness beyond their placebo effect in combatting male pattern baldness."[29] Even Dr. Schreck-Purola acknowledged that the medical community had reached such a conclusion. It has done so because it has found no credible theory explaining how these products work. As Dr. Ganiats explained, when, as in the case of the Helsinki Formula, "we can't imagine a reasonable mechanism of action," responsible scientists cannot conclude that the product is effective absent very strong evidence coming from well-designed studies. Dr. Kramer echoed this view, stating that "the standards to which you are held when one is testing an unorthodox theory really have to be quite rigid."

Pantron did not present any evidence that rebutted the consensus of the medical community that polysorbate-based products such as the Helsinki Formula are inherently ineffective. *All of the evidence of effectiveness adduced by Pantron can be explained by the placebo effect.*[30] Dr. Kramer offered uncontradicted testimony that hair growth studies reflect the existence of a very high placebo effect, as high as 41% in one study. Moreover, this placebo effect has an objective as well as a subjective component: not only do parties to the study misperceive hair growth, but patients will on occasion experience actual, measurable hair growth despite the fact that they have used a product of no intrinsic worth.[31]

None of Pantron's evidence of effectiveness takes the placebo effect into account. Pantron's evidence of consumer satisfaction is the most obviously flawed. The substantial placebo effect indicates that consumers simply cannot tell whether over-the-counter baldness cures are effective, inherently or otherwise. This is especially true in light of the irregular procession of hair loss — what an individual reports as the product's effectiveness in arresting hair loss may simply be the natural course of baldness. Much of Pantron's "consumer satisfaction" evidence

---

[29] Ibid. at 1091-1092.

[30] Emphasis is the author's.

[31] Although the reasons for this objective placebo effect are unclear, the testimony presented in the district court indicated that a likely explanation is the stimulation of the scalp that comes from massaging any product, including plain water, into the head on a regular basis. Id. at 1098.

is suspect on other grounds as well. Pantron's so-called "consumer satisfaction survey" was conducted by its own sales staff "as we did our follow ups to offer additional product." No record of the questions was kept. In addition, Pantron's low refund rate may not represent satisfaction. As Dr. Andreasen testified, even dissatisfied consumers may fail to exercise their right to a refund, because they think it not worth the trouble, because they feel guilty for having been deceived, because they credit the product's ineffectiveness to their own failure to follow instructions, or for any one of a number of other reasons.

Similarly, Pantron's clinical studies — and the expert testimony that relied solely on these studies — simply failed to account for the placebo effect. It is undisputed that these studies were not placebo-controlled. Pantron argues, however, that despite the lack of placebo controls, these studies were valid measures of the Helsinki Formula's effectiveness. First, it argues that Dr. Schreck-Purola's tests involved scalp biopsies that eliminated all subjectivity in the measurement of hair loss. Yet, because the study was neither controlled nor blinded, it could not account for the naturally irregular course of hair loss, nor for biased observation. Most significantly, it could not account for the objective aspect of the placebo effect.

Pantron also argues that "the French and Finnish studies were controlled by 'historical controls.'" The company contends that, because the participants in the study had previously tried many other remedies without success, the lack of results the participants had achieved in the past served as a control. Yet the designs of these studies never explicitly incorporated "historical controls," and they did not make a detailed comparison of the polysorbate-based products' results with the results the participants had achieved previously. Finally, Dr. Kramer offered unruffled testimony that historical controls are especially poorly suited for hair loss studies because of the irregular progression of male pattern baldness.

Pantron relies on Dr. Schreck-Purola's testimony that the success rate in the Schreck-Purola and Pons studies was too high to be explained by the placebo effect. Although the Rogaine studies showed a placebo effect of only 30–40%, the Schreck-Purola study showed hair growth in 60% of the subjects, and the Pons study showed hair growth in 80% of the subjects. Yet as Pantron's statistical expert conceded, it is improper to compare placebo rates across different studies, because "the placebo effect ... is entirely dependent upon the experimental design and the people doing the evaluation and the protocol that's been established." Absent a true control, Pantron's studies simply do not rebut the clinical and other scientific evidence presented by the FTC, which clearly demonstrates that any effectiveness of the Helsinki Formula arises solely from the placebo effect.

Assessing this evidence, the district court concluded that the FTC had failed to carry its burden of showing that the Helsinki Formula is "wholly ineffective." In essence, the district court held that, as a matter of law, a seller can represent that its product is effective even when this effectiveness is based solely on the placebo effect. We believe that the district court misapprehended the law.

As an initial matter, we should explain that we reject the argument vigorously urged by the FTC, that the district court clearly erred as a factual matter in determining that "the Helsinki Formula most probably works some of the time for a lot of people." The Commission argues that this finding is inconsistent with another finding made by the district court, that there was "no scientifically valid evidence that polysorbate 60 is effective for treatment of hair loss or for inducing growth." In essence, the FTC urges that we should hold that contemporary standards of scientific evaluation — which preclude the consideration of the placebo effect — are the determinants of what is "true" and what is "false." In its view, when the application of these contemporary scientific standards would lead to the conclusion that a product is ineffective, any claim that the product is effective is "false" in an intrinsic, absolute sense. We disagree. Contemporary scientific standards obviously are not the definitive or sole measure of what is "true" or "false." Galileo's theories were contrary to then-contemporary scientific standards, but we treat as a given that these theories were as essentially "true" when he explained them as they surely are today. Moreover, depending on our terms of reference, it may well not be incorrect to say that an efficacy representation is "true" when the product's effectiveness results solely from the placebo effect: for, in a certain sense, it would be "true" for a seller of sugar pills to represent that they relieve pain for some of the people some of the time, just as it would be "true" for

Pantron to state that the Helsinki Formula sometimes arrests hair loss. Whether because of psychological factors or because of the physiological effects of regularly massaging any product into the scalp, the evidence makes clear that the Helsinki Formula does work to some extent to combat baldness in some people some of the time.

However, neither scientific standards on the one hand, nor the broadest possible definition of "truth" on the other, can determine what constitutes a "false advertisement" under section 12 of the Federal Trade Commission Act. Indeed, a "false advertisement" need not even be "false"; it need only be "misleading in a material respect."[32] We must read this definition of "false advertis[ing]" in light of the overriding purpose of the FTC Act: "to protect the consumer from being misled by governing the conditions under which goods and services are advertised and sold to individual purchasers."[33] The question we must face, then, is not whether Pantron's claims were "true" in some abstract epistemological sense, nor even whether they could conceivably be described as "true" in ordinary parlance. Rather, we must determine whether or not efficacy representations based solely on the placebo effect are "misleading in a material respect," and hence prohibited as "false advertis[ing]" under the Act.

Taking account of these principles, we hold that the Federal Trade Commission is not required to prove that a product is "wholly ineffective" in order to carry its burden of showing that the seller's representations of product efficacy are "false." Where, as here, a product's effectiveness arises solely as a result of the placebo effect, a representation that the product is effective constitutes a "false advertisement" even though some consumers may experience positive results. In such circumstances, the efficacy claim "is 'misleading' because the [product] is not inherently effective, its results being attributable to the psychosomatic effect produced by the advertising and marketing of the [product],[34] as well as (in cases such as this one) the objective effects caused by the use of any product or even non-product in treating the condition in question. The court in *Acu-Dot*

---

[32] 15 U.S.C. §55.

[33] *National Petroleum Refiners Assoc. v. F.T.C.*, 482 F.2d 672, 685 (D.C.Cir. 1973), cert. denied, 415 U.S. 951 (1974); see also, *supra*, pages 1095-1097 (discussing the Cliffdale test).

[34] *U.S. v. An Article … Acu-Dot …*, 483 F. Supp. 1311, 1315 (N.D. Ohio 1980).

considered a magnetic patch, which the manufacturer had represented as effective in relieving muscle and joint pain. It concluded that "any therapeutic value of the [patch] is the result of its placebo effect,"[35] and accordingly held that the manufacturer's representations were "misleading" under 21 U.S.C. 352(a). In reasoning which fully applies to section 12 of the Federal Trade Commission Act, the court noted that "[a] kiss from mother on the affected area would serve just as well to relieve pain, if mother's kisses were marketed as effectively as the Acu-Dot device,"[36] and that a consumer purchasing a pain-reliever would expect it to have more therapeutic value than such a kiss.

The Acu-Dot court's reasoning is persuasive here. Under the evidence in the record before us, it appears that massaging vegetable oil on one's head would likely produce the same positive results as using the Helsinki Formula. All that might be required would be for Wesson Oil to remove Florence Henderson as its flack and substitute infomercials with Mr. Vaughn [the Pantron representative] that promote its product as a baldness cure. As the Commission has explained, the purposes of section 12 of the FTC Act dictate that a court should not allow a seller to rely on such a placebo effect in supporting a claim of effectiveness: "The Commission cannot accept as proof of a product's efficacy a psychological reaction stemming from a belief which, to a substantial degree, was caused by respondent's deceptions." Indeed, were we to hold otherwise, advertisers would be encouraged to foist unsubstantiated claims on an unsuspecting public in the hope that consumers would believe the ads and the claims would be self-fulfilling.[37]

Moreover, allowing advertisers to rely on the placebo effect would not only harm those individuals who were deceived; it would create a substantial economic cost as well, by allowing sellers to fleece large numbers of consumers who, unable to evaluate the efficacy of an inherently useless product, make repeat purchases of that product.[38]

The evidence before the district court made clear that there is no reason to believe that the Helsinki Formula is at all effective

[35] Ibid. at 1314.
[36] Ibid. at 1315.
[37] *Bristol-Myers Co.*, 102 F.T.C. 21, 336 (1983).
[38] See *Thompson Medical*, 104 F.T.C. at 718 (initial decision).

outside of its placebo effect. Accordingly, it was materially "misleading" under *Cliffdale Associates* for Pantron to represent that the Formula is effective in combatting male pattern baldness. We "resist the impulse to allow [Pantron] to market a product that works only by means of a placebo effect on the basis that it nevertheless often achieves a [result] as claimed."[39] Rather, we conclude that the district court erred in deciding that the FTC had not shown that Pantron's effectiveness claims were false.

In light of our conclusion, we instruct the district court to remove the portion of its injunction which allowed Pantron to "state that the Helsinki Formula is effective to some extent for some people." Such a representation — which, on the record before us, rests solely on the placebo effect — is misleading for the reasons set forth above. Moreover, we believe that the misleading nature of this statement is not cured by the district court's requirement that such a representation "be accompanied by clear and conspicuous disclosure that the product's effectiveness (1) is more likely to involve arrest of hair loss than growth of new hair, and (2) is not explained or supported by scientific studies recognized under standards in use in the U.S." The first of these limitations does not in any way detract from Pantron's claim — which, on the record before us, is "false" as a matter of law — that the Helsinki Formula is effective. It merely provides a more precise prediction of the manner in which the product will have its purportedly positive effects. As to the second limitation, it fails to provide full and fair information to the consumer and is therefore itself misleading. Scientific studies recognized under standards in use in the U.S. do not merely fail to explain or support Pantron's effectiveness claims; on the record before us, it is clear that they refute these claims and demonstrate that the Helsinki Formula has no effectiveness aside from its placebo effect. On remand, the district court shall modify its injunction to prohibit the company from making any representations that the Helsinki Formula is effective in arresting hair loss or promoting hair regrowth.

The district court shall also eliminate that portion of its current injunction which allows Pantron to "state that the Helsinki Formula (or a product similar thereto) was the subject of medical investigative work by responsible European physicians." In the context of an advertisement, such a statement carries the message

---

[39] *U.S. v. An Article ... Acu-Dot ...*, 483 F. Supp. at 1315.

that responsible scientific studies demonstrate that the Formula is effective. This statement is materially misleading. To be sure, the district court required the statement to be "accompanied by clear and conspicuous disclosure that the work did not conform to recognized standards in the U.S. for medical/scientific studies." However, we believe that this proviso, like the other proviso added by the district court, does not resolve the problem it seeks to address. A representation that responsible European studies demonstrate the Helsinki Formula's effectiveness is misleading not merely because these studies did not conform to U.S. scientific standards, but because all available evidence developed under the far higher American standards demonstrates the opposite of the European studies: that the Helsinki Formula is ineffective aside from its placebo effect. Pantron may not rely on the European studies in its advertisements, unless it discloses all facts necessary to ensure that the use of the study results is not misleading. If it wishes to cite the European studies, Pantron must disclose, at a minimum: (1) that recognized standards of medical and scientific experimentation in the U.S. are stricter than those under which the European studies were performed; (2) that researchers employing recognized American standards have studies the effectiveness of polysorbate-based hair growth products like the Helsinki Formula; and (3) that the unanimous conclusion of these researchers is that the Helsinki Formula and other similar products have no inherent curative or restorative effect.

13.3   Random Representative Samples

We showed in Chapters 1 and 2, how the courts have come to accept that samples can take the place of the entirety. *Pantron* further illustrates the courts' desire that all such samples be random and representative.

In Sections 1.2.3 and 3.3, we documented several cases in which the court objected to the survey methodology. Here is another. A psychiatrist was convicted of defrauding the Medicare Program and other insurance providers by billing for services he did not provide.[40] At sentencing, the government claimed the losses amounted to almost $1.2 million. This figure was obtained by extrapolating from the losses observed during the 19 months from September 1992 through March 1994 to the 6 years 1989 through 1994. The court refused:

---

[40] *U.S. v. Skodnek*, 933 F. Supp. 1108 (D. Mass., 1966).

to extrapolate from known losses in ways not supported by statistical theory, to cover a period outside that covered by the superseding indictment, based on flawed assumptions inadequately grounded in the trial evidence or sentencing data.[41] The data on which the government relies is skewed .... The government did not begin with a random sample. Instead it was a convenience sample, garnered by a unit whose purpose it is to investigate fraud. It was conducted not by dispassionate interviewers .... Clearly, the interviewers were searching out "horror" stories .... Indeed, there were instances in which reports apparently inconsistent with the overall conclusions were ignored.[42]

13.4   Power of a Test

We defined Type I error, rejecting a true hypothesis, and the probability of making such an error, the significance level, in Section 9.3.2. and introduced several methods for its calculation in Chapter 10. But what if we accept a null hypothesis when there really is an effect? This is a Type II error and could occur easily if the effect is small and the sample size not large enough. How can we demonstrate that a Type II error may have been made and how can we avoid the consequences of such an error?

Figures 13.1a amd 13.1b illustrate the close relationship between *power* (the probability of rejecting a hypothesis and accepting the alternative when the alternative is true), significance level, sample size, and the underlying difference between the hypothesis and the true alternative. Several basic principles may be abstracted from this figure:

1. The larger the underlying difference, the greater likelihood that we will detect it, that is, the greater the power.
2. The larger the sample, the greater the power.
3. If our sample cannot be enlarged (as is often the case when doing destructive sampling, see Section 1.1), we may be forced to compromise between the probabilities of making Type I and Type II errors, between significance level and power.[43]

---

[41] Id.

[42] Id.

[43] A word of caution: the relationship between significance level and power is much the same as the relationship in a paternity suit between the probability of exclusion and the likelihood of paternity. If the probability of exclusion with the tests employed is 95%, this means of 100 non-fathers, 95 will be excluded and 5 not excluded. This does not mean there is a 95% chance that the alleged father is the real father. As shown in Figure 13.1, power and significance level may be quite distinct.

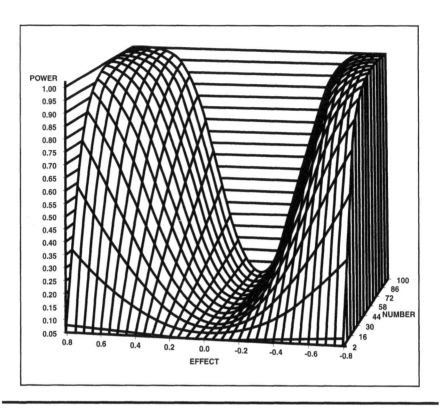

Figure 13.1a   Power of the two-tailed *t*-test with p = .05 as a function of the effect size (EFFECT) and the number of observations (NUMBER, $n_1 = n_2$) under the classical parametric assumptions.

Our first line of defense when told a null hypothesis must be accepted is to ask about the power of this test; see Goldstein [1985] and Zumbo and Hubley [1998]. One pointed example is the determination by some courts to ignore significant results if they can be reversed by swapping labels on one or two observations (see Section 9.3.1); a marked reduction in power is the inevitable consequence [Kadane, 1990].

13.4.1   Sample Size

Of the many rules for determining sample size considered by the courts,[44] most statisticians would agree that the *power of a test* should be the determining principle.

In *Pantron*,[45] a statistician testified that one of the studies was invalid because its sample size (68 men received polysorbate 60 and 73 men

---

[44] See Chapter 9.
[45] *FTC v. Pantron I Corp.*, 33 F.3d 1088 (9th Cir. 1994).

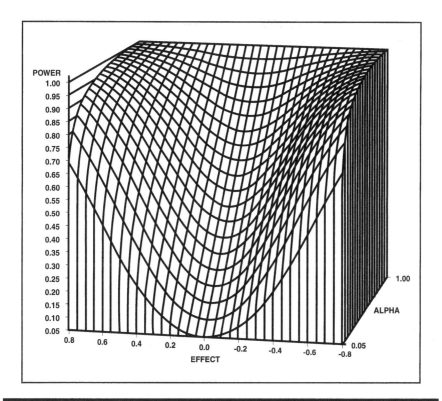

Figure 13.1b Power of the two-tailed *t*-test with sample sizes of $n_1 = n_2 = 20$ as a function of the effect size (EFFECT) and the significance level (ALPHA) under the classical parametric assumptions.

received placebos) was inadequate. He stated there was a 40% chance that the experiment would falsely conclude that an effective product is ineffective. He estimated that a minimum of 151 subjects in each group would be necessary for the study's conclusions to be reliable.

How can that be? As can be seen from Figures 13.1a nd 13.1b, the power of a test depends upon the magnitude of the effect as well as the sample size and the significance level. Supposing the placebo effect alone produces results 30% of the time. It would take a least a 10% increase in the chances of success, from 30% to 40%, to convince me to spend money for a baldness cure. At a significance level of 5%, an effect of this magnitude would be detectable at least half the time with two samples of size 50, and more than 60% of the time with two samples of the indicated size.

Perhaps Pantron's statistician had some other effect and some other significance level in mind. In reality, to every sample size corresponds an entire set of power curves.

In the preceding example, I would have been satisfied with a 10% increase in effectiveness. Would you insist on a higher percentage or be content with a lower number? What if the condition were not baldness, but a fatal disease (such as AIDs) that was virtually incurable without the drug?

13.4.2 Confidence Intervals

Too often, we report a single number, a point estimate, when we would be safer to report a range of possibilities.[46] One possibility is to make use of the *standard error* or variance of the estimate. Unfortunately, in most cases the standard error is only an approximation to the actual dispersion. If the observations are normally distributed (see Section 10.3.2) or if the sample is large, then the interval from one standard error below the sample mean to one standard error above will cover the true mean of the population about two-thirds of the time. However, if one of the other distributions depicted in that section more accurately portrays the underlying population, particularly one that is not symmetric such as the exponential distribution (Figure 10.1c), then the use of a symmetric interval makes little sense.

More generally, we can obtain a confidence interval for a given parameter by listing all the values of that parameter for which we would accept the hypothesis that it is the correct value. Suppose, for example, we observe that the sample median is 10. We find that a test of the hypothesis that the median is 11 would be accepted at the 5% significance level, but a hypothesis that the median is 11.1 would be rejected. Thus, the upper boundary of our confidence interval is 11. Similarly, at the lower end, we find that a hypothesis that the parameter is 9.4 would be accepted, but the hypothesis that the parameter is 9.3 would be rejected. We describe the interval from 9.4 to 11 as a 95% confidence interval for the parameter.

We could obtain tighter limits, say from 9.6 to 10.7, by using a 90% significance level, but then we could only be confident that our interval using this statistical approach would include the correct value of the parameter 90% of the time.

If you are trying to argue that a certain drug has a damaging effect, but the confidence interval for the size of that effect includes zero, your case is lost (at least on those grounds).

A detailed description of how to derive confidence intervals is given in Good [2000; Chapter 5].

Silicone Breast Prostheses

The following case, a seeming exception to the rule that silicone implants are bad per se, illustrates the successful use of a defense based on the use of confidence intervals and an attack on the method of analysis.

---

[46] If the observations are measured on a continuous scale, from, say, 1.0000 to 16.0000 the probability of observing any specific value is infinitesmal.

An action was brought against the manufacturer of Heyer-Schulte breast implants, in which plaintiff claimed her receipt of two of the manufacturer's breast implants in 1977 had caused her to develop Sjögren's syndrome, an inflammatory disorder with symptoms of dry eyes, dry mouth, and dry vagina. The district court excluded proposed expert testimony of the plaintiff's epidemiologist and rheumatologist as to the issue of whether breast implants can cause Sjögren's syndrome or its symptoms, and granted the manufacturer's motion for judgment as matter of law on the issue of causation. The rationale for the exclusion was that the epidemiologist relied on a study that yielded a lower-end confidence bound of less than one for the relative risk linking breast implants to Sjögren's syndrome.

The epidemiologist's reanalysis of the epidemiological study was also inadmissible to establish causation, since a one-tailed statistical test had been employed during reanalysis rather than the preferred two-tailed test, even though the epidemiologist had not seen the data collected in the study. Her reanalysis of study data was unpublished, not generally accepted, and contained within it the *a priori* assumption that breast implants have negative health effects on women, which introduced an unknown and potentially devastating amount of error into reanalysis.

Since the epidemiologist could not testify regarding a causal link between breast implants and Sjögren's syndrome, she also could not testify regarding a possible causal link between breast implants and symptoms of Sjögren's syndrome. The proposed testimony of plaintiff's rheumatologist regarding general causation was inadmissible because studies on which the rheumatologist sought to rely were not reasonably supportive of the proposition that breast implants cause Sjögren's syndrome. The rheumatologist acknowledged that his own studies had not been published or peer reviewed, his sampling technique was fraught with bias, his theory regarding general causation was not generally accepted in the medical community, and he had not tested his theories. In the absence of any evidence regarding a general causal link between breast implants and Sjögren's syndrome, the rheumatologist was not permitted to testify as to specific causation. that is, that the plaintiff's implants had caused her to suffer from Sjögren's syndrome. Moreover, even if breast implants could cause Sjögren's syndrome, the rheumatologist's testimony on specific causation would be excludable since it failed the test of parsimony and did not meet standards under *Daubert*, insofar as he did not base his diagnosis on a wide reading of the medical literature but, instead, based it on studies that did not support the finding of a general causal link between implants and Sjögren's syndrome.[47]

---

[47] *Kelley v. American Heyer-Schulte Corp.*, 957 F Supp. 873, (W.D. Tex. 1999); 46 Fed. Rules Evid. Serv. 1359.

Audits

The Office of Audit Services of the U.S. Department of Health and Human Services distributes via the Worldwide Web a computer program, RAT-STATS, for assistance in selecting (stratified) random samples and determining the appropriate sample sizes. In particular:

> This program allows the user to estimate the sample size for a given precision [interval length] at a given confidence level. Sample size estimates are based on the assumption that a variable appraisal will be performed.

> The program will generate sample sizes for unrestricted and stratified samples. The user may also enter the overall sample size for a stratified sample and the program will determine the optimum allocation among the strata. The user needs to enter the mean, universe and standard deviation for each stratum in order to generate the estimated sample sizes. The program allows for sample sizes for up to 12 strata.[48]

13.5 Coincidence and the Law of Small Numbers

Imagine that you are on retainer to a small hospital that has recently experienced an unfortunate number of "incidents." As a result it faces an administrative hearing later this month and the possibility of a temporary or permanent closing of the facility. At a meeting called to discuss possible remedies, you notice that all eyes turn toward Dr. Singh, the cardiologist, who sits at the far end of the table.

Remedies are discussed and you even suggest a few of your own, all of which are shot down either because they are already in place or because — like closing the emergency room and accepting only less critical patients — they are illegal. During the break, you pick up a newspaper and scan the baseball statistics. In an instant, you formulate the hospital's defense — it is a small hospital, after all.

Table 13.1 reproduces the statistics as of April 21, 1999 for the Anaheim Angels. Note that the team has several .400 hitters, which seems incredible since no one has ever equaled Ted Williams' 1963 season. In fact, all of the batters are hitting great although few fans expect to see the Angels in the playoffs this year.

What is the explanation? It is the same as the one for a small hospital that performs only a limited number of critical procedures each month.

---

[48] *RAT-STATS User Guide*, p. 198.

Table 13.1 Angels Averages as of April 16, 1999

Batter	AVG	AB	R	H	2B	3B	HR	RBI	BB	SO
Glaus	.462	39	11	18	10	0	2	10	4	5
Salmon	.419	43	12	18	5	0	4	15	8	9
Verlarde	.354	48	8	17	2	1	0	4	2	11
Sheets	.350	40	7	14	4	0	1	4	2	10
Anderson	.304	46	8	14	1	0	1	7	2	7
Walbeck	.292	24	3	7	0	0	0	1	1	5
Erstad	.289	45	7	13	4	0	0	5	6	6
Greene	.263	38	5	10	2	0	3	13	2	8
Palmeiro	.256	39	7	10	1	0	1	4	0	3

When a sample is small, the number of events (the AB column in Table 13.1) and their averages can be markedly different from the values anticipated in the long run.

Right fielder Tim Salmon has a career average of .296; despite the impressive number of hits he has produced in his initial 43 times at bat (a .462 average!). The Angels' manager anticipates Tim will achieve about a .296 average again this year once he has accumulated sufficient at-bats.

Similarly, Dr. Singh has an impressive lifetime record of success despite the difficulties inherent in his specialty. The two fatalities that occurred last month are of far less significance when looked at in the context of a lifetime. While none of this absolves the hospital of the need to look closely at its current surgical procedures, it provides a cogent and, hopefully, convincing explanation of why a small hospital can and will experience short periods during which its incident rates are markedly different from the desired levels.

13.5.1 Clustering

Where chance alone rules, clustering, not uniformity, is the norm. Two fatalities occur in a single month following a 2-year period that was accident free. Three witnesses to the Kennedy assassination die in the same year. Chance? Or sinister forces at work?

To see how chance alone might be responsible, consider a simple example that can be worked easily with pencil and paper. Suppose four incidents are observed over a one-year period. Label the events A, B, C, and D and draw a chart of their possible occurrences by season as in Table 13.2.

This table will have a total of 256 rows. If you do not want to spend the time completing the table, consider that each of the four incidents might

Table 13.2   Distribution of Incidents by Season

Spring	Summer	Fall	Winter
A	B	C	D
A	C	B	D
AB	C	D	
AC	B	D	
A	CB	D	
			ABCD
...	...	...	...

have occurred in any of the four seasons resulting in a total of 4 × 4 × 4 × 4 = 256 possibilities.

The probability that each incident occurred during a different season is given by the corresponding number of rows (24) divided by the total number of rows, that is 24/256 < 10%. The probability that all four incidents occurred in at most two quarters of the year is 84/256, a little less than one third of the time.

13.5.2   The Ballot Theorem and the Arc–Sine Law

Fortunately, for the non-mathematically inclined reader, we will not attempt to prove or even enunciate either of the obtuse mathematical results referenced in the title to this section. What we will do is discuss their implications.

> Contrary to generally accepted views, the laws governing a prolonged series of individual observations will show patterns and averages far removed from those derived for a whole population. In other words, currently popular psychological tests would lead one to say that in a population of "normal" coins most individual coins are "maladjusted."[49]

Once you get behind in a situation where chance is the predominant factor, it will take you an unexpectedly long time before you catch up. Suppose, for example, that hospital regulations specify that no more than one identifiable "error" is permitted for each ten surgical procedures performed, and by chance the very first set of ten results in two errors. How soon will things even out? If the error rate is indeed one in ten,

---

[49] Feller [1968; p. 72].

then the probability is only 35% that the correction will take place in the next set of 10; worse, the probability is 25% that two or more errors will be observed which would only exacerbate the situation. With the next inspection only 6 months away, will the hospital be ready on time? As noted above, the answer is "probably not." If you can't catch up, you may be able to at least stay in place. This time, the probabilities will be on your side. What one can and should argue, particularly if the hospital has instituted a set of remedial procedures, is that the incidence rate during the latter half of the inspection period is satisfactory even if the incidence rate during the initial half was not.[50]

13.6 Coincidence and Ad Hoc–Post Hoc Arguments

Anything can and will happen in the long run. Misleading probabilities, lack of reproducibility, and clusters of events are only a few reasons why ad hoc–post hoc arguments should be rejected out of hand.

13.6.1 Reproducibility

No reputable scientist would ever report his results before he succeeds in reproducing his findings at least twice, once in his own laboratory and once in that of a colleague. An outside test can be particularly telling as all too often some overlooked factor — such as the quality of the laboratory water — is responsible for the results, *not* the factors under investigation. It is better to be found wrong in private than in public. Reproduce a result, then reproduce it again.

13.6.2 Painting the Bull's Eye around the Bullet Holes

Perci Diaconis [1978] spent some years as a statistician investigating paranormal phenomena including those linked to Uri Geller, the man who claimed he could bend spoons with his mind. Diaconis was disappointed but not surprised to find that the hidden "powers" of the occult were more or less those of the average nightclub magician, down to and including forcing a card and taking advantage of ad hoc–post hoc hypotheses. The fact is, as anyone who has played poker will concede, that one out of every two hands contains "something" interesting. The magician's "trick" lies in saying, "Look at this! Isn't this incredible?"

---

[50] We are assuming the initial run of bad luck was only that, bad luck.

In Section 13.5.1, we examined the occurrences of "incidents" over the course of four seasons and showed that chance alone could easily be responsible for their uneven distribution throughout the course of the year. Our implicit assumption was that the nature of the incident — misdiagnosis, pilot error, or assembly line failure — had been determined in advance. When rock stars or political figures die, we tend to define the nature of the incident after the fact.

When three buses appear at your stop simultaneously, a stand of cherry trees is found amid a forest of oaks, six cases of leukemia occur in a small town, and the night sky viewed through a telescope is filled with clusters of stars, do you holler "Magic" or remember the Poisson distribution that we studied in Section 10.3.3? The best courtroom defense when your opponent paints a bull's eye around the bullet holes is to provide a computer simulation producing precisely the same results by chance alone (see Freedman [1983] for an excellent example).

13.6.3   Data Mining or Searching for Significance

When we perform a statistical test at a 5% significance level, what we are really saying is that, at least once out of each 20 tests (5%), a result that is not significant will appear significant purely by chance. Consider the Baldus study that we examined in Sections 12.4 and 12.7.3. The number of variables, 253, was truly impressive, but a total of 253 variables means that 5% or 13 of them will have statistically significant correlations by chance alone.

The solution, again, is to avoid ad hoc–post hoc hypotheses. Scientists normally work in sequential fashion, taking several small exploratory samples initially, as they develop their hypotheses, and then one or two large samples when they proceed to test them. In the case where we have limited data at hand, as in an audit for scientific integrity, one possibility is to divide the data into two parts, then use the first part to help formulate the hypotheses and the second part to test them. Only in this fashion can we be sure that the variables that test successfully for significance are not simply the results of coincidence.

13.7   Bad Statistics

Primary attacks on statistical results should be and will be directed against the design of the experiment or survey. Consider the factors listed in Chapter 8: bias in selection and response, confounding, observer bias, and inconsistent classifications. Was the population appropriate or germane (see Chapter 1)? What about those who did not respond? Would their responses be the same as those who did?[51]

---

[51] Nonresponders include those who refused to answer or participate as well as those who were not available or were simply overlooked.

Attacks can and will be directed against the methods of collection as they are today in the ongoing review of the decennial census. Does an audit trail allow one to go backward from the numbers stored in the computer to the original observations? At the very least, a random selection (audit) should be made to determine the number and extent of the errors in transcription.

13.7.1   An Example from the National Game

We include the next case because it reiterates important points raised in many of the preceding chapters and, to be honest, because one of the parties is Bud Selig, the current Major League Commissioner; and I am a rabid baseball fan.[52]

Selig v. U.S.[53] pitted the commissioner against the Internal Revenue Service (IRS) which disallowed his proportionate share of the amortizations of certain player contracts that the then newly formed Milwaukee Brewers acquired from the ashes of the Seattle Pilots. A two-step regression analysis of player salaries was offered in evidence by Selig and disputed by the IRS.

Though the Court stated, "I decline to resolve this dispute,"[54] nonetheless, it rejected the regression analysis on the following grounds, all of which we have encountered at least once in previous chapters:

■ Wrong population — transaction data from the player market was used as the basis of the regression equation, a market that is highly controlled by the American League, whereas "the relevant market is the club market in which the bundle of assets was purchased."[55]

■ Sample not representative — the sample consisted of transactions in the player market, whereas "testimony was unanimous that higher quality players are rarely transacted for cash in the player market."[56]

■ Wrong predictor variables — the amount by which the player reserve system depressed salary levels was erroneously attributed to the value of the franchise.

■ Database not reliable — both parties agreed at trial that the transaction records lacked much information about the true substance of the transactions.

---

[52] My daughters and I sat side by side to watch Nolan Ryan, pitching for the Astros, leave our beloved Cubbies hitless in nine.

[53] 565 F. Supp. 524 (E.D. Wis. 1983).

[54] Id. at 539.

[55] Ibid.

[56] Ibid. at 540.

- R^2 too small to be of predictive value — the sample size of 36 was inadequate given that 13 coefficients in the regression equation had to be determined.
- Wrong variable predicted — the theoretical salary predicted by the equations consistently underestimated the actual contract values.

Question the Data[57]

Is the measurement process reliable?
Is the measurement process valid?
Were the data recorded correctly?
Were the units observed the units at issue?

Question the Design

Were the observations independent?
What are the confounding variables?
Can the results be generalized?

Question the Analysis

Were all test assumptions satisfied?
Was the most powerful test used?

Question the Presentation

Do charts and graphs portray data fairly?
Are rates and percentages properly interpreted?
Were appropriate measures of location, precision, and association used?

13.7.2   Large Sample Approximations

The majority of statistics in use today represents asymptotic (large-sample) approximations whose use predates the ready availability of personal computers. The chi-square statistic and the F statistic, for example, are valid only for very large samples and, as important, only when each of the subsamples or factor divisions is also sizeable (incorporating ten observations or more).

For example, an analysis of the data in Table 13.3 derived from a study of oral lesions yields a far-from-significant p value of 14% when the chi-

---

[57] See Kaye and Freedman [1994] for further detail.

Table 13.3 Oral Lesions in Three Regions of India

Site of Lesion	Kerala	Gujarat	Andhra Pradesh
Labial Mucosa	0	1	0
Buccal Mucosa	8	1	8
Commissure	0	1	0
Gingiva	0	1	0
Hard Palate	0	1	0
Soft Palate	0	1	0
Tongue	0	1	0
Floor of Mouth	1	0	1
Alveolar Ridge	1	0	1

square statistic is employed, whereas the true significance level as determined by permutation methods is 2%.

The burden of proof is on the proposer of a given statistic to demonstrate that all underlying assumptions are satisfied. For example, to apply an F statistic to the analysis of variance, the individual data items must be (1) independent of one another, (2) identically distributed, and (3) taken from a normal distribution. Counter-examples abound. Here are just a few:

■ Repeated measurements on the same individual are not independent.
■ Successive prices of a stock are not independent, while day-to-day changes may be.
■ Volunteers differ from the average, more passive, participant; thus, the first mouse selected from a cage tends to be more active, more aggressive, and have higher corticosteroid levels than the last mouse selected.

13.7.3 Multiple Statistics, Multiple Conclusions

In some situations, a choice of statistics is available. For example, in a two-sample comparison one might use either a t test or a permutation test. The two tests yield roughly equal results in the long run, accepting and rejecting in the same proportion of cases, but in any specific case, they may yield quite different results. It is important to determine during discovery whether the opposing statistician has gone "shopping" for the most favorable result.[58]

---

[58] Chapter 15 discusses questions to ask during discovery.

13.8 Counterattack

Four factors considered in earlier chapters apply to any statistical technique:

- Data must be complete and accurate.
- Appropriate populations must be used for comparison purposes.
- Samples sizes must be adequate.
- Assumptions underlying the statistical model must be satisfied.

More precise measurements almost always yield more powerful tests. For example, a test that is based on the actual values of a variable will require far fewer samples for a given power and significance level than one for which our only knowledge is whether or not its value exceeded a certain critical threshold.

Do not accept single-valued (point) estimates. Insist on being provided with confidence intervals.

Use distribution-free statistics in preference to those that are distribution-dependent; fewer assumptions provide fewer grounds for challenge.

A result based on many factors is superior to one based on only a few; the courts recognize that a multivariate regression provides results superior to a contingency table or a correlation coefficient.

Use one statistical expert to offset the reports of another. In many cases, the courts will simply step away from the controversy.

Question the accuracy and precision of the data, the appropriateness of the survey or experimental design, and the validity and appropriateness of the analysis and of the presentation. Timely objections must be made at trial.

Chapter 14

What Every Statistician Should Know about Courtroom Procedure

The graduations and ranks of the courts are infinite, extending beyond the ken even of initiates. The proceedings in the courts of law are generally a mystery to the lower officials as well; therefore, they can almost never follow the progress of the cases they are working on throughout their course; the case enters their field of vision, often they know not whence, and continues on, they know not where.[1]

Kafka thoroughly understood the trial process. Consider, Joseph K.'s experiences in Kafka's *The Trial*. First, came a lengthy process of behind-the-scenes activity in which the authorities decided whether a trial would be appropriate and the defendant cast about for possible lines of defense. Then came an equally lengthy process of discovery in which various essential documents were elicited from the opposing parties and subjected to intense scrutiny. Witnesses were made to undergo lengthy interrogations while the pretrial activity ground on. The trial was over in a few short days and statisticians who may have labored for years on trial preparation were not invited ins.de. In a small fraction of real-life cases, regrettably not in the case of Joseph K., a trial could be followed by an appeal.

---

[1] Somewhat modified from Kafka's *The Trial*.

In an informal poll of the statisticians I know who have done legal work I found that 30% worked with attorneys prior to filing, 70% participated at some point during the discovery process though only 35% actually gave depositions, 10% testified during trials, and 14% worked on appeals. We review the statistician's role during each of these phases in what follows.

14.1 Selecting the Case

You may want to decline some or all of the cases you are offered because you find the client morally objectionable or the case is without merit, or merely to avoid the embarrassment and humiliation to which an expert witness may be subjected during a hotly contested trial.

The U.S. system of justice is adversarial. You are not asked to serve in a dispassionate search for justice, but are expected to represent your client's interests to the fullest. Once you accept a case, professional ethics require that you present only one point of view and prepare substantive arguments rebutting other views. While the fees you can demand from attorneys are near the top of the scale, you should turn down any clients you feel you would be unable to represent adequately.

This raises an ethical issue. Should you as a professional be prepared to see all sides of an argument? The answer is yes until you form your initial opinion and agree to accept ongoing payments from an attorney. Your initial decision should not be made lightly or without a thorough, careful review of all the evidence at hand. Should you later alter your opinion as a result of additional evidence or additional insight, you should immediately notify the attorney who engaged you, but should this additional insight come to you suddenly on the second day of a brutal cross-examination by an opposing attorney who bears an uncommon resemblance to the schoolyard bully, your only ethical outlet is tears or feigned insanity. Your best defense against bullies is thorough, careful preparation.

14.2 Prefiling

Prior to filing, the statistician serving the plaintiff can assist in answering three questions:

> Does the data support the charge?
> What is the magnitude of any as-yet-unrealized damages?
> What additional data is needed to answer the preceding questions?

The statistician for the defense, assuming he or she has been well briefed by his or her attorney, will address these same issues and, anticipating the answers of his or her opposite number, will be asked to outline lines of defense.

You may be asked to collect and analyze the data. If so, make sure to anticipate any and all objections to the collection process. Favor the simple analysis (and the simple explanation) over the complex. The sensitivity analysis described in Section 9.3.1 is an excellent example.

14.3   Discovery

The objectives of discovery are threefold:

- To get all the relevant facts before the court
- To encourage settlement prior to trial
- To streamline the trial process by committing each side to particular facts and procedures

Statisticians will agree with the American Bar Association that the "need for full and fair disclosure is especially apparent with respect to scientific proof and the testimony of experts." This sort of evidence is practically impossible to test or rebut at trial without the opportunity to examine it closely.[2]

"There are no satisfactory grounds for withholding information in the discovery process."[3] If you are to be called as a witness, virtually every facet of your career is open to discovery (including any unpleasant or best-overlooked items). If you have merely aided the attorney in the organization and presentation of the case, examining and commenting on documents, then everything you have done is privileged, considered part of the attorney's own work product, and not subject to examination.

My best advice in the face of these conflicting rulings is to keep your notes "barebones," but not so skimpy that you will be unable to flesh them out at trial should you be called to testify.

14.4   Depositions

As part of the discovery process, the opposing attorney may take the opportunity to question you prior to trial, usually in his office, but possibly on neutral ground. (Yes, you will get paid, but by the other attorney.)

---

[2] *ABA Standards Relating to Discovery and Procedures before Trial*, 66.
[3] *State v. Tankersley*, 191 Ariz. 359, 956 P.2d 486, 495 (1998).

Both the time and the place of the deposition should be at least partially at your convenience.

Whether testifying before trial during the taking of a deposition or testifying at trial, you have one primary guideline: keep it simple. This does not mean that the statistics must be simple. It means that your well-rehearsed explanation should be straightforward and coherent.

The object of the deposition on the attorney's part is to probe his opponent's case for weaknesses. Your own objective while being deposed is to reveal as little as possible while conforming to the rules of discovery.

You may be tempted to stray from your rehearsed opinion, particularly if you feel that your honor is at stake or your judgment is questioned. Do not stray. The opposing attorney who will examine you is not your friend, will never be your friend, and cannot be won over to your side, however reasonable your arguments are.

The opposing attorney's questions will focus on the weaknesses rather than the strengths of your position.[4] Your best defense is to keep your answers brief and, if pressed, to repeat the arguments you have already made. Never volunteer information. Whether or not the opposition has employed a statistician, the worst thing you as an expert witness can do is to volunteer a defense for an objection that has not been raised.

Listen to the questions carefully. Answer as briefly as possible. Take your time. You are not on a quiz show; you get no bonus points for the quickest or even the best answer. Do not qualify your answers or provide analogies or counter-examples. If you think of something that might clarify your answer, suppress the thought.

You do not need an answer for every question. It is better to be thought a fool than to open your mouth and prove you are one. Very likely, you will be asked, "What other methodologies were considered and why were they rejected?" The best answer you can give, at least from your own attorney's point of view, is "none." Do not lie; answer "none" only if this is truly the case. If a better answer occurs to you later in the interrogation, suppress it; take it home and tell it to your dog. To paraphrase warnings stated on cop shows, your answers are being taken down in evidence and will be used against you.

Do not guess. You may be asked for your opinions of various statistical methodologies, for example, Good's Test for Type I censored data. Your best answer is, "I am not familiar with a test under that name."[5]

---

[4] You should know; you prepared the list of nasty questions for your attorney to ask the opposing party's statistician.

[5] Of course if you're one of the few familiar with that rather obscure contribution to the statistical literature, then fire away.

In summary, your goal throughout the deposition should be the same as Dorothy's in Oz — to go home as quickly as possible where there are people who love you and care for you.

14.5 Post-Deposition, Pretrial Activities

Your attorney will expect you to prepare questions for use in his depositions of the opposing party's statistician, along with a list of documents to be elicited during discovery.[6] Afterward, you will be asked to review the responses and prepare a further list of questions to use at trial. Your responsibilities extend from the design of forms, through the collection and storage of data, to the actual analysis. You may need to recommend further analyses or the gathering of additional data to advance your client's case.

14.6 In the Courtroom

Your appearance at trial will have three phases:

1. You will sit for hours waiting to be called as a witness only to be sent home — again — without testifying.
2. You will undergo direct examination by the attorney who engaged you or one of his or her colleagues.
3. You will be cross-examined by the same attorneys who did their best to humiliate you during your deposition.

The rules for cross-examination are essentially the same as those that guided you during the taking of your deposition. Listen to the questions. Do not be quick to respond. When you respond, be truthful, brief, and to the point. Do not embellish.

Apart from the continuing admonishment to keep it simple, the rules for direct testimony during trial are quite different from the rules for a deposition. During direct testimony, you will be doing your best to instruct a judge and jury. Teach — use all the tricks you learned in the classroom to engage the listeners. Use analogies. Will visual aids be more effective? Your attorney can arrange to have an easel or a projector and screen brought into the courtroom so that you can present tables and graphs you prepared earlier. After displaying a graph or a pie chart, start listing main points on the easel (until and unless you are asked to sit down).

---

[6] See Section 15.3.1 for some suggestions.

Emphasize the positive. Talk about the procedures you used and your contribution to the analysis. Even if the data was a mess and you went through hell to clean it up and make it usable, a judge does not need to know that. If data had to be discarded, be prepared to account for it briefly and simply. Do not make bad data the focal point of your testimony.[7]

14.7 Appeals

Your work during appeals is much the same as it was during prefiling and pretrial: to recognize and comment on bad data and faulty analyses.

14.8 Summary

The statistician can make effective contributions at all stages of a trial by focusing on three issues:

Does the data support the charge?
What is the magnitude of any as-yet-unrealized damages?
What additional data is needed to answer the preceding questions?

Whether testifying before trial during the taking of a deposition or testifying at trial, the statistician has one primary guideline: keep it simple. This does not mean that the statistics themselves need be simple. It means that your well-rehearsed explanation must be straightforward and coherent.

The opposing attorney who will be examining you is not your friend, will never be your friend, and cannot be won over to your side, however reasonable your arguments. His questions will focus on the weaknesses rather than the strengths of your position. Be cool. Remember, when it's all over, you'll have your doctorate — oops! That was an entirely different ordeal.

---

[7] Unless you are testifying concerning the other statistician's data.

Chapter 15

Making Effective Use of Statistics and Statisticians

For the lawyers — and even the least important of them has at least a partial overview of the circumstances — are far from wishing to introduce or carry out any sort of improvement in the court system, while — and this is quite characteristic — almost every statistician, even the most simple-minded among them, starts thinking up suggestions for improvement from the moment the trial starts, and in doing so often wastes time and energy that would be better spent in other ways.[1]

In what follows, we consider the contributions a statistician can make at various points in the trial process, the likely areas of ignorance, and some guidelines to ensure a statistician will prove a help, not a hindrance.

15.1 Selecting a Statistician

You need a statistician if:

- The word *sample* is used.
- You need to conduct a survey or an experiment or appraise the results of a survey or an experiment.
- The opposing side has hired a statistician.

---

[1] Somewhat modified from Kafka's *The Trial*.

You should be looking for a statistician who:

■ Is a great communicator.
■ Possesses a breadth of knowledge and experience.

A good communicator can both explain and listen. He or she can explain statistical concepts so that you and your paralegal can understand them. He or she is willing to accept criticism and suggestions from you for improvement, can keep cool during depositions, and knows when to stop talking. A good communicator will listen to and follow the other person's arguments before attempting to rebut them.

Breadth of knowledge is more important than depth; you want someone who can find the holes in his or her own arguments as well as the opposition's, and who is not so wedded to techniques (the penalty sometimes of too much depth) as to be unable to utilize a statistical test or estimator more appropriate to the task at hand.

The statistician's credentials are not as critical as they are with other expert witnesses — because the statistician's explanations, not the statistician, will fall under scrutiny. A doctorate is probably preferable to a master's degree, and a doctorate in mathematical statistics coupled with some knowledge of the subject area is preferable to a doctorate in the subject coupled with some knowledge of statistics.

You can benefit by employing a statistician to supplement and interpret the testimony of other expert witnesses. In a trial pivoting on the analysis of DNA data, for example, consider employing both an expert on DNA analysis[2] and a statistician to provide significance levels for the expert's findings.

15.2 Prefiling Preparation

Most attorneys fail to make adequate use of statisticians during the prefiling process. A statistician can assist in estimating the extent of recoverable damages and assessing the validity of the evidence. He or she can dictate what further evidence must be collected and suggest possible lines of counterattack.

Recently, I worked for an attorney whose client was undergoing a government audit. A large fine for the client was the likely outcome. I sketched a possible line of defense based on possible differences among the various lines of business the client was engaged in. A single sample

---

[2] When judges weigh one expert's opinion against another's, credentials are essential; see, for example, *People v. Axell*, 235 Cal. App. 3d 836.

would not do because the error rates uncovered at audit would surely vary from product line to product line.[3] At the attorney's request, the client's staff set about gathering the supporting evidence needed to demonstrate the differences. The government would either have to broaden its inquiry — an expensive process — or limit its claim for damages to the business lines actually sampled.

Statisticians can fulfill similar roles during the discovery process, but only assuming it is not already too late for them to be effective.

15.3 Discovery

> Witnesses who understand the discovery and deposition process, who appreciate the differences in goals between depositions and courtroom testimony, and who are well versed in the techniques for effectively answering questions, survive this process without creating pitfalls to overcome at trial.[4]

A statistician is needed during the discovery process to specify what reports and documents are required, to comment on the completeness and adequacy of the documents retrieved, to help formulate a set of questions, and to aid in interpreting the responses.

Normally, any interchange, oral or written, with your statistician is exempt from discovery. The exception to the work product rule comes when the expert testifies as a witness.[5] If you anticipate that your statistician may testify, he or she should be admonished to limit the extent of what he or she commits to paper.

I aroused the ire of my attorney client (and rightly so) when I failed to heed his admonishment to provide oral reports only. (In my defense, I only committed my points to writing when I began to feel my oral reports were being tuned out. As with any other expert witness, a little attention to the statistician's findings on the attorney's part goes a long way to ensure a positive relationship.)

15.3.1 Questions

Here is a list of questions to use during discovery. Your statistician can help with interpretation and with rephrasing questions for recalcitrant witnesses.

---

[3] See Section 3.2.

[4] Starr [1998, p. 121].

[5] *U.S. v. Nobles*, 422 U.S. 225, 239 (1975).

Documents

Be sure to obtain copies of all reports and have your statistician examine them for completeness. The ideal report will permit the reader to replicate the study in its entirety. Those aspects of reports that fall short of this ideal warrant further inquiry.

The Witness (Credentials and Experience)

Jurists have made it clear that the credentials of an expert witness are important to them and help determine both admissibility and the weight to be given the evidence.[6] As always, avoid asking questions if you are not going to like the answers.

> What is your educational background?
> Is statistics your primary occupation?
> Have you participated in studies like this in the past?
> Get the details of the expert's experience. Have you previously determined sample sizes?
> Have you previously supervised the collection and storage of data?
> How did you acquire your knowledge of the specific statistical techniques employed in this case?

The Sample

> How were individuals or items chosen for inclusion in the sample?
> Was a random mechanism employed?
> What was the nature of the mechanism?
> Was the sample simple random? Clustered? Stratified?
> Were there controls?
> What was their nature?
> Were the controls adequate?
> In your expert opinion was the final sample representative of the population from which it was drawn?
> How were the sample and subsample sizes determined?
> In your expert opinion and in retrospect, are these sample sizes adequate?

Data Collection Forms

Obtain copies of all forms used. Ask whether these forms were pretested. If they were, ask for copies of all pretest results and forms. (The opposing

---

[6] See, for example, *People v. Axell supra.*

party may have kept changing the survey form or the experimental design until they got the results they wanted.)

Data Collection

How was the data collected?

How did you ensure the individual results were independent of one another?

What were the backgrounds of the individuals who collected the data? What training did the data collectors receive?

Was the training the same for all the data collectors?

How did you ensure the prescribed methods were actually followed?

Were all answers entered as given or was any interpretation of answers made prior to entry? (Get details of any interpretations.)

What percentage of forms was validated by resampling?

Who conducted the resampling?[7]

These questions are essential as the majority of errors is introduced during the collection process.[8] If the statistician being deposed denies any knowledge of the collection process, confirm and document the lack of knowledge as it can be quite revealing at trial.

What was the percentage of non-responders? (You may want to distinguish those who were simply unavailable from those who refused to answer.)

What attempts if any were made to reduce the number of non-responders?

Were the non-responders subsampled by other means?

Data Entry

How was the data entered into the computer?

Was its correctness validated on entry?

If so, how?

What percentage of responses was validated?

---

[7] In *Rust Environment & Infrastructure, Inc. v. Teunissen*, 131 F.3d 1210, 1218 (7th Cir. 1997), the court criticized a survey in part because it "did not comport with accepted practice for independent validation of the results."

[8] The census is a prime example; see Section 3.2.2.

Data Storage

How was the data stored?

What procedures did you employ to ensure the integrity of the data?

When did you last verify the integrity of the database?

Data Analysis

What statistical techniques did you employ? (Obtain coherent written descriptions of all techniques and formulas employed along with their justification.)

What was the power of your tests?[9]

What assumptions underlie these procedures?

Does the data have to be distributed in a certain way?

Do the observations need to be independent?

Did you verify your assumptions were satisfied?

What alternative statistical techniques did you consider?

What are the alternative techniques?

The Witness (Trial Experience)

Have you testified in other trials?

How many?

What was their nature?

Did you employ similar data collection and statistical procedures in those trials?

What occasioned the differences in the present case?

15.3.2 Depositions

Many statisticians (and other expert witnesses) consider themselves more intelligent than others. They labor under the illusion that once they have explained their interpretation of the data, the other side, awestruck by their brilliance, will pack it in and slink away. The best approach is to give them some indication of the mixture of indifference and antagonism that actually lies ahead, though, inevitably, they won't quite believe it until it happens.

---

[9] This question has a dual purpose. The concept of power, central to a formal study of statistical theory, is often a foreign concept to those who have only a passing knowledge of statistics. A blank look on the face of the so-called expert suggests this question and the expert's lack of knowledge need be followed up at trial.

On the other hand, your statistician can assist you is in reviewing the depositions of those who are collecting and analyzing the data for your opponent. Your statistician can prepare a list of questions for trial, and, if warranted, perform further analyses and/or gather additional data.

15.4 Presentation of Evidence

The questions in Section 15.3.1 and those in the sidebar included with Section 13.6. should form the basis of both your examination and cross-examination. Hopefully, your statistician will have provided you with a list of questions specific to the case at hand.

If you do not understand your statistician's presentation, neither will the judge and jury. In that situation, have him or her read Chapter 14 and try again.

Not all statisticians understand the need for visual aids and know how to use them effectively, but all statisticians now have access to statistics computer programs that can generate outstanding charts and graphs. Show your statistician examples of the types of graphic aids you require and he or she will be able to generate them for you.

As with other expert witnesses, statisticians can testify only to their methods and findings and a confidence interval or significance level. They cannot testify to any conclusions about other aspects of the trial. In *Mahan v. Farmers Union Central Exchange Inc.*,[10] for example, the court ruled:

> Statisticians may testify that their statistical tests show or do not show patterns of discrimination based on age, but may not testify to the ultimate conclusion that discrimination was or was not exercised [in the case at hand]. The jury should be the final arbiter of that issue.

In *State v. Jackson*,[11] the court ruled the trial judge erred in allowing the expert to state a conclusion that the defendant was "probably" the father of the victim's child.

15.5 Appeals

The material in this and preceding chapters has indicated the bases on which appeals may be filed on statistical grounds. Lynda Axell appealed

---

[10] 235 Mont. 410, 786 P.2d 850, 857 (1989).

[11] 320 N.C. 452 (1987).

from a judgment of conviction of first degree and attempted robbery, challenging the trial court's ruling on the following grounds:

- The DNA typing evidence failed to meet the *Kelly/Frye* rule.
- Frequency of the use of a novel technique and its admission elsewhere at trial was allowed to substitute for an analysis of its general acceptance.
- Weaknesses were present both in procedures followed and in statistical calculations.
- The testing laboratory failed to obtain adequate control values.

15.6 Summary

The statistician can be of invaluable assistance during all phases of a trial. He or she can assist you prior to filing by assessing the likelihood of success and the nature and extent of any additional data that may be required. He or she can assist you during discovery in evaluating documentation and formulating questions. The effectiveness of a statistician at trial may depend as much upon his or her skill as a communicator as his or her knowledge of statistics. A statistician can also provide invaluable service by interpreting and assigning statistical significance to the testimony of other expert witnesses.

A statistician also can be the source of many problems; there are many possibilities for misrepresentation on both sides. Keep the communication lines open and treat him or her with the care you would expend on any other expert witness.

References

Aitken CCG (1995) *Statistics and the Evaluation of Evidence for Forensic Scientists.* Chichester:Wiley.

Angell M (1996) *Science on Trial: The Clash of Medical Evidence and the Law.* New York:Norton.

Anscombe FJ (1973) Graphs in statistical analysis. *Amer. Statist.* 27:17.

Babcock BA (1975) *Voir Dire.* Preserving its wonderful power. *Stan. Law Rev.* 27:545.

Babcock BA (1993) A place in the palladium: women's rights and jury service. *U. Cinn. Law Rev.* 61:1139–1180.

Ball H (1986) *Justice Downwind: America's Atomic Testing Program in the 1950's.* Oxford University Press.

Barnes DW (1983) *Statistics as Proof.* Boston:Little Brown.

Beale S (1984) Integrating statistical evidence and legal theories in challenges to the selection of grand and petit juries. *Law Contemp. Prob.* 46:269.

Becker SJ (1991) Public opinion polls and surveys as evidence: suggestions for resolving confusion and conflicting standards governing weight and admissibility. *Or. Law Rev.* 70:463.

Berry DA (1991) Inference using DNA profiling in forensic identification and paternity cases. *Statist. Sci.* 6:175–206.

Berry DA; Geisser S (1986) Probability of paternity. In *The Use of Statistics in Forensic Science*, Aitken CGG; Stoney DA, Eds. Chichester:Ellis Horwood, 150–156.

Bierig J (1998) Methodological challenges to government sampling techniques. *Health Care Fraud Abuse Newsletter.* 1:7:1.

Boswell MT; Gore SD; Patel GP; Tallie C (1993) The art of computer generation of random variables. In *Handbook of Statistics*, 9, *Computational Statistics*, edited by Rao CR. Amsterdam:North Holland.

Broderick RJ (1992) Why the peremptory challenge should be abolished. *Temple Law Rev.* 369.

Carp RA (1982) Federal grandjuries: how true a cross-section of the community? *Just. Syst. J.* 7:257–277.

Cecil JS; Willging TE (1993) Court-Appointed Experts: Defining the Role of Experts Appointed under Federal Rule of Evidence 706, Washington, DC:Federal Judicial Center.

Chemerinsky E (1994) Civil rights: important decisions … but not dramatic. *Res Ipsa* 1:10–12.

Cleveland WS (1985) *The Elements of Graphing Data.* Monterey, CA:Wadsworth.

Cleveland WS (1993) *Visualizing Data.* Summit, NJ:Hobart Press.

Cochran WG (1977) *Sampling Techniques* (3rd ed.). New York:Wiley.

Coleman RF; Walls HJ (1974) The evaluation of scientific evidence. *Crim. Law Rev.* 276–287.

Converse JM; Presser S (1986) *Survey Questions: Handcrafting the Standardized Questionnaire.* Newbury Park, CA:Sage.

Cullison AD (1969) Probability analysis of judicial fact finding: a preliminary outline of the subjective approach. *U. Toledo. Law Rev.* 1969:538–598.

Dant M (1988) Gambling on the truth: the use of purely statistical evidence as a basis for civil liability.

Davis KC (1971) *Discretionary Justice.* Chicago:University of Illinois.

Diaconis P (1978) Statistical problems in ESP research. *Science* 201:131–136.

DiPrima SR (1995) Selecting a jury in federal criminal trials after Batson and McCollum. *Colum. Law Rev.* 95:888–911.

Duncan AJ (1986) *Quality Control and Industrial Statistics* (5th ed.). Homewood, IL:Irwin.

Dutka S (1982) The use of survey research in legal proceedings. *ABA J.* 68:1508–1510.

Evidence: admission of mathematical probability statistics held erroneous for want of demonstration of validity. (1967) *Duke Law J.* 665:675–678.

Farley; Mosteller F (1979) A conversation about Collins. *U. Chi. Law Rev.* 41:242.

Federal Judicial Center (1982) *Manual for Complex Litigation* ¶2.712, at 118.

Feller W (1968) Introduction to Probability Theory and its Applications, 3rd ed. New York:Wiley.

Fienberg SE (1993) *The Evolving Role of Statistical Assessments as Evidence in the Courts.* New York:Springer-Verlag.

Fink A; Kosecoff B (1998) *How to Conduct Surveys: A Step by Step Guide.* Newbury Park, CA:Sage.

Finkelstein MO; Farley WB (1970) A Bayesian approach to identification evidence. *Harv. Law Rev.* 83:489.

Finkelstein MO; Farley WB (1971) A comment on "trial by mathematics." *Harv. Law Rev.* 84:1801–1809.

Finkelstein J (1973) The application of statistical decision theory to the jury discrimination cases. *Harv. Law Rev.* 80:338.

Fisher (1980) Multiple regression in legal proceedings, *Colum. Law Rev.* 80:702.

Fisher RA (1934) The logic of inductive inference (with discussion). *J. Roy. Statist. Soc. A.* 98:39–54.

Freedman DA (1983) A note on screening regression equations. *Amer. Statist.* 37:152–155.

Freedman DA (1994) Adjusting the census of 1990. *Jurimetrics J.* 34:99–106.

Freedman DA (1999) From association to causation: some remarks on the history of statistics. *Stat. Sci.* 14:243.

Fukurai H; Butler EW; Krooth R (1991) Cross-sectional jury representation of systematic jury representation? Simple random and cluster sampling strategies in jury selection. *J. Crim. Just.* 19:31–48.

Gastwirth JL (1988) *Statistical Reasoning in Law and Public Policy.* Academic Press.

Gastwirth JL (2000) *Statistical Science in the Courtroom.* New York:Springer.

Gallant AR (1987) *Nonlinear Statistical Models.* New York:Wiley.

Goldstein R (1972) Interdistrict inequalities in school financing: a critical analysis of *Serrano v. Priest* and its progeny, *U. Pa. Law Rev.* 120:504, 423–525, nn. 67, 71.

Goldstein R (1985) Two types of statistical errors in employment discrimination cases. *Jurimetrics J.* 26:32–47.

Good PI (2000) *Permutation Tests* (2nd ed.). New York:Springer-Verlag.

Good PI (2001) *Resampling Methods.* (2nd ed.) Boston:Birkhauser.

Hamilton HG (1998) The movement from *Frye* to *Daubert*: Where do the states stand? *Jurimetrics J.* 38:201–213.

Harr J (1991) *A Civil Action.* New York:Random House.

Harrison J (1990) Peremptory challenges and the meaning of jury representation. *Yale Law J.* 89:1177–1198.

Hill AB (1971) *Principles of Medical Statistics* 9th ed. New York:Oxford University Press.

Kadane JB (1990) Statistical analysis of adverse impact of employer decisions. *J. Amer. Statist. Assoc.* 85:925.

Kalven H; Zeisel H (1966) *The American Jury.* Chicago:University of Chicago Press, 498.

Kaplan (1968) Decision theory and the fact-finding process. *Stan. Law Rev.* 20:1066.

Karys M (1977) Jury representativeness: a mandate for multiple source lists. *Ca. Law Rev.* 65:776–798.

Kaye DH (1986) Ruminations or jurimetrics: hypergeometric confusion in the fourth circuit. *Jurimetrics J.* 26:215–223.

Kaye DH (1986) Is proof of statistical significance relevant? *Wash. Law Rev.* 61:1333.

Kaye DH (1988) *Plemel* as a primer on proving paternity. *Willamette Law Rev.* 24:867.

Kaye DH (1989) The probability of an ultimate issue: the strange case of paternity testing. *Iowa Law Rev.* 75:75.

Kaye DH (1998) Bible reading: DNA evidence in Arizona. *Az. State Law J.* 28:1035–1077.

Keynes JM (1921) *A Treatise on Probability.* London:Macmillan.

Klein SP; Freedman DA (1993) Ecological regression in voting rights cases. Chance. 6:38–43.

Kleinbaum DG; Kupper LL; Muller KE (1988) *Applied Regression Analysis and Other Multivariate Methods.* Boston:PWS-Kent.

Kougasian PM (1993) Should judges consider the demographics of the jury pool in deciding change of venue applications? *Fordham Urban Law J.* 20:531–550.

LaFave W; Israel J (1984) *Crim. Proc.* 2:13.2(a), 160.

Lempert R (1991) Some caveats concerning DNA as criminal identification evidence with thanks to the Reverend Bayes. *Cardoza Law Rev.* 13:303–341.

Lempert R (1994) Suspect population and DNA identification. *Jurimetrics J.* 34:1–7.

Levin B; Robbins H (1983) Urn models for regression analysis with application to employment discrimination studies. *Law Contemp. Prob.* 46:246–267.

Meier P; Sacks J; Zabell SL (1984) What happened in Hazelwood; statistics, employment, discrimination and the 80% rule. *Amer. Bar Found. Res. J.* 139–186.

Meier P; Zabill S (1980) Benjamin Peirce and the Howland will. *J. Amer. Statist. Assoc.* 75:497–506.

Monahan J; Walker L (1985) *Social Science in Law.* Foundation Press, 211–212.

Mosteller F; Tukey JW (1977) *Data Analysis and Regression.* Reading, MA:Addison-Wesley.

Nietzel MT and Dillehay RC (1986) *Psychological Consultation in the Courtroom.* New York:Elsevier Science.

Note (1974) Height standards in police employment and the question of sex discrimination. *S. Cal. Law Rev.* 47:585.

Note (1966) Mathematical probabilities misapplied to circumstantial evidence. *Minn. Law Rev.* 50:745.

Paetzold RL; Willborn SL (1994 and updated annually) *The Statistics of Discriminology: Using Statistical Evidence in Discrimination Cases.* New York:Shepard's/McGraw-Hill.

Parascandola M (1998) What is wrong with the probability of causation? *Jurimetrics J.* 39:29–44.

Peterson R (1982) A few things you should know about paternity tests (but were afraid to ask). *Santa Clara Law Rev.* 22:667.

Rabinovitch NL (1969) Studies in the history and probability of statistics XXII: probability in the Talmud. *Biometrika* 56:437–441.

Redmayne M (1998) Bayesianism and Proof. In *Science in Court*, Freeman M; Reece H (eds.). Brookfield MA:Ashgate.

Rubinfeld DL, Ed. (1991) Statistical and demographic issues underlying voting rights cases. *Eval. Rev.* 15:659.

Shroeder YC (1987) The procedural and ethical ramifications of pretesting survey questions. *Amer. J. Trial Advocacy* 11:195–201.

Shoben A (1978) Differential pass-fail rates in employment testing: statistical proof under Title VII. *Harv. Law Rev.* 91:793–811.

Smith RL; Charrow RP (1975) Upper and lower bounds for the probability of guilt based on circumstantial evidence. *J. Amer. Statist. Assoc.* 70:555–560.

Starr VH (1998) *Witness Preparation.* New York:Aspen.

Starr VH; Jordan WE (1993) *Jury Selection: An Attorney's Guide to Jury Law and Methods.* Boston:Little, Brown.

Tribe L (1971) Trial by mathematics: precision and ritual in the legal process. *Harv. Law Rev.* 84:1329.

VanDyke JM (1977) *Jury Selection Procedures.* Cambridge, MA:Ballinger.

von Mises R (1928, 1957) *Probability, Statistics, and Truth.* London:MacMillan.

Whittaker J (1990) *Graphical Models in Applied Statistics.* Chichester:Wiley.

Ylvisaker D (1986) Comment: on blood test reports in paternity cases. In *Statistics and the Law*, Degroot M; Fienberg S; Kadan J (eds.). New York:John Wiley, 383–390.

Zeisel H; Kaye D (1997) *Prove It with Figures: Empirical Methods in Law and Litigation*. New York:Springer-Verlag.

Zumbo BD and Hubley AM (1998) A note on misconceptions concerning prospective and retrospective power. *Statistician* 47:385–388.

Table of Authorities

Accord, Robinson v. City of Dallas, 514 F.2d 1271 (5th Cir. 1975).

Adams v. Superior Court of Los Angeles County, 27 Cal. App. 3d 719, 104 Cal. Rptr. 144 (1972).

Albemarle Paper Co. v. Moody, 422 U.S. 405 (1975).

Aldasoro v. Kennerson, 922 F. Supp. 339 (S.D. Cal. 1995).

Alexander v. Louisiana, 405 U.S. 625 (1972).

Allen et al. v. U.S., 588 F. Supp. 247 (1984).

American Basketball Ass'n v. AMF Voit, Inc., 358 F. Supp. 981 (S.D. N.Y.), aff'd, 487 F.2d 1393 (2nd Cir. 1973), cert. denied, 416 U.S. 986 (1974).

American Fed'n of State, County, and Municipal Employees, AFL-CIO v. Washington, 770 F.2d 1401 (9th Cir.1985).

American Iron & Steel Institute v. EPA, 115 F.3d 979 (D.C. Cir. 1997).

Amstar Corp. v. Domino's Pizza, Inc., 205 U.S.P.Q 128 (N.D. Ga. 1979), rev'd, 615 F. 2d 252 (5th Cir. 1980).

Andersen & Co. v. U.S., 284 F.542 (9th Cir. 1922).

Anderson v. Bessemer City, 470 U.S. 564 (1985).

Anderson v. Liberty Lobby, Inc., 477 U.S. 242 (1986).

Andrews v. State, 533 So.2d 841 (Fla. App. 1988).

Appalachian Power Co. v. EPA, 135 F.3d 791 (D.C. Cir. 1998).

Arlington Heights v. Metropolitan Housing Dev. Corp., 429 U.S. 252 (1977).

Attorney General v. Irish People, Inc., 684 F.2d 928 (D.C. Cir. 1982), cert. denied, 459 U.S. 1172 (1983).

Ballard v. U.S., 329 U.S. 186 (1946).

Ballew v. Georgia, 435 U.S. 223 (1978).

Barnes v. GenCorp Inc., 896 F.2d 1457 (6th Cir), cert. denied, 498 U.S. 878 (1990).

Barnes v. Glen Theatre, Inc., 501 U.S. 560 (1991).

Basko v. Sterling Drug, Inc., 416 F.2d 417 (2nd Cir. 1969).

Batson v. Kentucky, 476 U.S. 79 (1986).

Bazemore v. Friday, 751 F.2d 662 (4th Cir. 1984), aff'd in part, vacated in part, remanded, 478 U.S. 385 (1986).

Berry v. Chaplin, 74 Cal. App. 2d 652 (1946).

Bethlehem Steel Co. v. Industrial Accident Comm'n., 21 Cal.2d 742, 8 Compensation Cases 61 (1943).

B.F. Goodrich Co. v. Dept. of Transportation, 541 F.2d 1178 (6th Cir. 1976).

Blake v. City of Los Angeles, 595 F.2d 1367 (9th Cir. 1979).

Bodenschatz v. State Personnel Board, 15 Cal. App. 3d 775, 781.

Boston Chapter NAACP v. Beecher, 504 F.2d 1017 (1st Cir. 1974).

Bouman v. Block, 940 F.2d 1211 (9th Cir. 1991).

Bridgeman v. Commonwealth, 3 Va. App. 523 (1986).

Bridgeport Guardians Inc. v. Members of Bridgeport Civil Service Comm'n., 354 F. Supp. 778 (D. Ct.), modified 482 F.2d 1333 (2nd Cir. 1973), cert. denied, 421 U.S. 991 (1975).

Bristol-Myers Co., 102 F.T.C. 21 (1983).

Brock v. Merrell Dow Pharmaceuticals, Inc., 874 F.2d 307 (5th Cir.), modified on reh'g, 884 F.2d 166 (5th Cir. 1989), cert. denied, 494 U.S. 1046 (1990).

Brocklehurst v. PPG Industries, 123 F.3d 890 (6th Cir. 1997).

Bush v. Kentucky, 107 U.S. 110 (1883).

Capaci v. Katz & Besthold Inc., 525 F. Supp. 317 (E.D. La. 1981).

Caldwell v. State, 260 Ga. 278 (1990).

Carter v. Texas, 177 U.S. 442 (1900).

Carter v. Jury Comm'n, 396 U.S. 320 (1970).

Castaneda v. Partida, 430 U.S. 482 (1977).

Celotex Corp. v. Catrett, 477 U.S. 317 (1986).

Certified Color Manufacturers Ass'n. v. Mathews, 543 F.2d 284 (D.C. Cir. 1976).

Chaves County Home Health Serv., Inc. v. Sullivan, 931 F.2d 914 (D.C.Cir. 1991), cert. denied, 502 U.S. 1091 (1992).

Chemical Mfrs. Ass'n v. EPA, 870 F.2d. 177 (5th Cir. 1989).

Chemical Mfrs. Ass'n v. EPA, 28 F.3d 1259 (D.C. Cir. 1994).

Chicano Police Officers Association v. Stover, 526 F.2d 431 (10th Cir. 1975).

Chlorine Chemistry Council v. EPA, (D.C. Cir. 2000) No. 98-1627.

Christy v. Hodel, 857 F.2d 1324 (9th Cir. 1988), cert. denied, 490 U.S. 1114 (1989).

Chumbler v. Commonwealth, 905 SW.2d 488 (Ky. 1995).

City of New York v. Dept. of Commerce, 822 F. Supp. 906 (E.D. N.Y. 1993).

Clady v. County of Los Angeles, 770 F.2d 1421 (9th Cir. 1985).

Clinchfield R.R. Company v. Lynch, 527 F. Supp. 784 (E.D. N.C. 1981), aff'd, 700 F.2d 126 (4th Cir. 1983).

Coble v. Hot Springs School District No. 6, 682 F.2d 721 (8th Cir. 1982).

Cole v. Cole, 74 N.C. App. 247, aff'd, 314 N.C. 660 (1985).

Commonwealth v. Beausoleil, 397 Mass. 206 (1986).

Commonwealth v. Curnin, 409 Mass. 218 (1991).

Commonwealth of Pennsylvania et al. v. Rizzo et al., 466 F. Supp. 1219 (E.D. Pa. 1979)

Commonwealth of Pennsylvania v. Local Union 542, Int'l Union of Operating Engineers, 469 F. Supp. 329 (E.D. Pa. 1978), aff'd, 648 F.2d 9222 (3rd Cir. 1981), rev'd sub nom, *General Building Contractors Ass'n., Inc. v. Pennsylvania*, 102 S. Ct. 3141 (1982).

Coleman v. Prudential Relocation, 975 F. Supp. 234 (W.D. N.Y. 1997).

Connecticut v. Teal, 457 U.S. 440 (1982).

Georgia v. McCollum, 505 U.S. 42 (1992).

Glasser v. U.S., 315 U.S. 60 (1942).

Glaxo, Inc. and Glaxo Group Limited v. Novopharm Ltd., 110 F.3d 1562, 42 U.S.P.D. 2d 1257 (4th Cir. 1997).

Gomillion v. Lightfoot, 364 U.S. 339 (1960).

Gonzalez v. State, 643 SW.2d 751 (Tex. App. 4th Dist. 1982).

Gregg v. Georgia, 428 U.S. 153 (1976).

Grier v. Kizer, 219 Cal. App. 3d 422, 268 Cal. Rptr. 244 (1990).

Griffith v. State of Texas, 976 S.W.2d 241 (1998).

Griggs v. Duke Power Co., 401 U.S. 424 (1971).

Guenther v. Armstrong Rubber Co., 406 F.2d 1315 (3rd Cir. 1969).

Gulf South Insulation v. Consumer Product Safety Commission, 701 F.2d 1137 (5th Cir. 1983).

Hall v. E.I. DuPont de Nemours & Co., Inc., 345 F. Supp. 353 (E.D. N.Y. 1972).

Harper v. Trans World Airlines, Inc., 525 F.2d 409 (8th Cir. 1975).

Haskell v. Kaman Corp., 743 F.2d 113 (2nd Cir. 1984).

Hawley Products Co. v. U.S. Trunk Co., 259 F.2d 69 (1st Cir. 1958).

Hazelwood School District v. U.S., 392 F. Supp. 1276 (E.D. Mo.), rev'd, 534 F.2d 805, vacated and remanded, 433 U.S. 299 (1977).

Hazen Paper Co. v. Biggins, 507 U.S. 604 (1993).

HCA Health Services v. Kansas, 900 P.2d 838 (Kan. Ct. App. 1994).

Hernandez v. Texas, 347 U.S. 475 (1954).

Hopkins v. Dow Corning Corp., 33 F.3d 1116 (9th Cir. 1994).

Hose v. Chicago and Northwestern Transp. Co., 70 F.3d 968 (8th Cir. 1995).

Houston v. Lafayette County, 56 F.3d 606 (5th Cir. 1995).

Illinois Physicians Union v. Miller, 675 F.2d 151 (7th Cir. 1982).

In Re Agent Orange Product Liability Litigation, 597 F. Supp. 740 (E.D. N.Y. 1984), aff'd, 818 F.2d 145 (2nd Cir. 1987), cert. denied, 484 U.S. 1004 (1988)).

In Re Forte-Fairbairn, Inc., 62 F.T.C. 1146 (1963).

In Re Joint Eastern & Southern District Asbestos Litigation, 827 F. Supp. 1014 (S.D. N.Y. 1993).

In Re Joint Eastern & Southern District Asbestos Litigation, 964 F.2d 92 (2nd Cir. 1992).

In Re Joint Eastern & Southern District Asbestos Litigation, 758 F. Supp. 199 (S.D. N.Y. 1991).

In Re Joint Eastern & Southern District Asbestos Litigation, 52 F.3d 1124 (2nd Cir. 1995).

J.E.B. v. Alabama ex rel. T.B., 511 U.S. 127 (1994).

Jackson v. Nassau County Civil Service Comm'n., 24 F. Supp. 1162 (E.D. N.Y. 1976).

Jenkins v. Red Clay Consol. School District Board of Educ., 4 F.3d 1103 (3rd Cir. 1993), cert. denied, 512 U.S. 1252 (1994).

Johnson v. Miller, 864 F. Supp. 1354 (S.D. Ga. 1994), aff'd, 515 U.S. 900 (1995).

Johnson v. Transportation Agency, 480 U.S. 616 (1987).

Kaplan v. Internat'l Alliance of Theatrical Artists, 525 F.2d 1354 (9th Cir. 1975).

Kelley v. American Heyer-Schulte Corp., 957 F. Supp. 873 (W.D. Tex., 1997).

Kelley v. Federal Energy Regulatory Comm'n., 96 F.3d 1482 (D.C. Cir. 1996).

Kie v. U.S., 27 F. 351 (C.C. Ore. 1886).

King's Case, 352 Mass. 488, 225 N.E.2d 900 (1967).

Koger v. Reno, 98 F.3d 631 (D.C. Cir. 1996).

Penney v. Praxair, Inc., 116 F.3d 330 (8th Cir. 1997).

People v. Brown, 40 Cal.3d 512 (1985).

People v. Castro, 545 N.Y.S.2d 985 (1989).

People v. Coleman, 46 Cal.3d 749 (1988).

People v. Collins, 68 Cal.2d 319, 66 Cal. Rptr. 497, 438 P.2d 33, 36 ALR3d 1176 (1968).

People v. Harbold, 124 Ill. App. 3d 363 (1st Dist 1984), 79 Ill. Dec. 830, 464 NE.2d 734.

People v. Risley, 214 N.Y. 75 (1915), 108 N.E. 200.

People v. Estrada, 93 Cal. App. 3d 76 (1979), 55 Cal. Rptr. 731.

People v. Fields, 35 Cal.3d 329, 673 P.2d 680, 197 Cal. Rptr. 803 (1983).

People v. Flores, 62 Cal. App. 3d Supp. 19, 133 Cal. Rptr. 759 (1976).

People v. Harmon, 215 Cal. App. 3d 552, 263 Cal. Rptr. 623 (1989).

People v. Harris, 36 Cal.3d 36, 201 Cal. Rptr. 782 (1984), cert. denied 469 U.S. 965, appeal to remand, 191 Cal. App. 3d 819, 236 Cal. Rptr. 680, appeal after remand, 217 Cal. App. 3d 1332, 236 Cal. Rptr. 563.

People v. Johnson, 47 Cal.3d 1194, 767 P.2d 1047, 255 Cal. Rptr. 569 (1989).

People v. Jones, 9 Cal.3d 546, 510 P.2d 705, 108 Cal. Rptr. 345 (1973).

People v. Lesara, 206 Cal. App. 3d 1305, 254 Cal. Rptr. 417 (1988).

People v. Manson, 71 Cal. App. 3d 1, 139 Cal. Rptr. 275 (1977), cert. denied, 435 U.S. 953 (1978).

People v. Motton, 39 Cal.3d 596, 217 Cal. Rptr. 771 (1985).

People v. Pride, 3 Cal.4d 195, 10 Cal. Rptr. 2d 636 (1992).

People v. Sirhan, 7 Cal.3d 710, 102 Cal. Rptr. 385 (1972), cert. denied, 410 U.S. 947.

People v. Slone, 76 Cal. App. 3d 611.

People v. Stansbury, 4 Cal.4th 1017, 17 Cal. Rptr. 2d 174 (1993).

People v. Taylor, 113 Ill.3d 467, 447 N.E.2d 519 (1983).

People v. Tevino, 39 Cal.3d 667, 225 Cal. Rptr. 652 (1985).

People v. Viscotti, 2 Cal.4th 1, 5 Cal. Rptr. 2d 495 (1992).

People v. Wheeler, 22 Cal.3d 258, 148 Cal. Rptr. 890 (1978).

People v. White, 43 Cal.2d 740, 278 P.2d 9 (1954).

People v. Yorba, 209 Cal. App. 3d 1017 (1989).

Peters v. Kiff, 407 U.S. 493 (1972).

Pittsburgh Press Club v. U.S., 579 F.2d 751 (3rd Cir. 1978).

Pollis v. New School for Social Research, 132 F.3d 115 (2nd Cir. 1996).

Powers v. Ohio, 499 U.S. 400 (1991).

Protestant Memorial Medical Center, Inc. v. Dept. of Public Aid, 295 Ill. App. 3d 249 (1998).

Puerto Rico Maritime Shipping Authority v. Federal Maritime Comm'n., 678 F.2d 327 (D.C. Cir. 1982).

Pulley v. Harris, 465 U.S. 37 (1984).

Ratanasen v. California Dept. of Health Serv., 11 F.3d 1467 (9th Cir. 1993).

Reynolds v. Sheet Metal Workers Local 102, 498 F. Supp. 952 (D.D.C. 1980).

Rhodes Pharmacal Co. v. FTC, 208 F.2d 382 (7th Cir. 1953), rev'd in part on other grounds, 348 U.S. 940 (1955).

Ristaino v. Ross, 424 U.S. 589 (1976).

Robinson v. Mandell, 20 F. Cas. 1027 (C.C.D. Mass., 1868).

Romero v. City of Pomona, 665 F. Supp. 853 (C.D. Cal. 1987), aff'd, 883 F.2d 1418 (9th Cir. 1989).

Rose v. Mitchell, 443 U.S. 545 (1979).

Rust Environment & Infrastructure, Inc. v. Teunissen, 131 F.3d 1210 (7th Cir. 1997).

Samuels v. Air Transport Local 504, 992 F.2d 12 (2nd Cir. 1993).

San Antonio Independent School District v. Rodriguez, 411 U.S. 1 (1973).

Sawyer v. U.S., 148 F. Supp. 877 (M.D. Ga. 1956).

Scott v. Perini, 622 F.2d 428 (6th Cir. 1981), cert. denied, 456 U.S. 909 (1982).

Segar v. Civiletti, 508 F. Supp. 690 (D.D.C. 1981).

Segar v. Smith, 738 F.2d 1249 (D.C. Cir. 1984), cert. denied, 471 U.S. 1115 (1985).

Sengupta v. Morrison-Knudson Company, Inc., 804 F.2d 1072 (9th Cir. 1986).

Sheehan v. Daily Racing Form, Inc., 104 F.3d 940 (7th Cir. 1997).

Sheehan v. Purolator, Inc., 839 F.2d 99 (2nd Cir. 1988).

Sierra Club v. Costle, 657 F.2d 298 (D.C. Cir. 1981).

Simblest v. Maynard, 427 F.2d 1 (2nd Cir. 1970).

Simpson v. Midland-Ross Corp., 823 F.2d 937 (6th Cir. 1987).

Sindell v. Abbott Laboratories, 26 Cal.3d 588, (1980).

Smith et al. v. Virginia Commonwealth University, 84 F.3d 672 (4th Cir. 1996).

Smith v. Rapid Transit Inc., 317 Mass. 469, 58 N.E.2d 754 (1975).

Smith v. Salt River, 109 F.3d 586 (9th Cir. 1997).

Smith v. Texas, 311 U.S. 128 (1941).

Spellen v. Allen, 344 U.S. 443 (1953).

Spencer v. Commonwealth, 238 Va. 275 (1989).

Stewart v. State, 256 Ga. 70, 268 S.E. 906 (1980).

Strader v. West Virginia, 100 U.S. 303 (1880).

Sobel v. Yeshiva University, 566 F. Supp. 1166 (S.D. N.Y. 1983), 839 F.2d 18 (1983).

State v. Boles, 905 P.2d 572 (Ariz. Ct. App. 1995).

State v. Garrison, 585 P.2d 563 (Ariz. 1978).

State of Georgia Dept. of Human Resources v. Califano, 446 F. Supp. 404 (N.D. Ga. 1977).

State v. Clark, 887 P.2d 572, 164 Ariz. Adv. Rep. 68 (1994).

State v. Garrison, 120 Ariz. 255, 585 P.2d 563 (1978).

State v. Hernandez, 192 Wis.2d 251, 531 NW.2d 348 (1995).

State v. Jackson, 320 N.C. 452, 358 S.E.2d 679 (1987).

State v. Schweitzer, 533 N.W.2d 156 (S.D. 1995).

State v. Sneed, 76 N.M. 349, 414 P.2d 858 (1966).

State v. Passino, No. 185-1-90 Fcr (Dist Ct. Franklin County, May 13, 1991).

State v. Pennington, 393 S.E.2d 847 (1990).

State v. Spann, 130 N.J. 484 (1993).

Stephens v. State, 456 S.E.2d 560 (Ga. 1995).

Strauder v. West Virginia, 100 U.S. 303 (1880).

Summers v. Tice, 33 Cal.2d 80, (1948).

Taylor v. Louisiana, 419 U.S. 522 (1975).

Teague v. Attala County, 92 F.3d 283 (5th Cir. 1996).

Teamsters v. U.S., 431 U.S. 324 (1977).

Texas Dept. of Community Affairs v. Burdine, 450 U.S. 248 (1981).

Thiel v. Southern Pacific Co., 328 U.S. 217 (1946).

Tinker v. Sears, Roebuck & Co., 127 F.3d 519 (6th Cir. 1997).

Thornburg v. Gingles, 478 U.S.30 (1986).

Thompson v. Saatachi Holdings, 958 F. Supp. 808 (W.D. N.Y. 1997).

Thompson Medical, 104 F.T.C. at 718 (initial decision).

Toldeo, St. Louis & Western R.R. Co. v. How, 191 F. 776 (6th Cir. 1911).

Tomka v. Seiler Corp., 66 F.3d 1295 (2nd Cir. 1995).

Toys "R" Us, Inc. v. Canarsie Kiddie Shop, Inc., 559 F. Supp. 1189 (E.D. N.Y. 1983).

Turner v. Fouche, 396 U.S. 346 (1970).

Turner v. Murray, 476 U.S. 28 (1986).

Turpin v. Merrell Dow Pharmaceuticals, Inc., 959 F.2d 1349 (6th Cir.), cert. denied, 506 U.S. 8264 (1992).

U.S. v. 43 1/2 Gross Rubber Prophylactics, 65 F. Supp. 534 (Minn. 4th Div. 1946).

U.S. v. Bailey, 862 F. Supp. 277 (D. Colo. 1994), aff'd in part, rev'd in part, 76 F.3d 320, cert. denied, 116 S. Ct. 1889.

U.S. Dept. of Labor v. Harris Trust and Savings Bank, No. 78-OFCCP-2 (ALJ decision, Dec. 22, 1986).

U.S. Postal Service Board of Governors v. Aikens, 460 U.S. 711 (1983).

U.S. v. Gwaltney, 790 F.2d 1378 (9th Cir. 1986), 20 Fed. Rules Evid. Serv. 1293, cert. denied, 479 U.S. 1104.

U.S. v. Aguilar, 883 F.2d 662 (9th Cir. 1989), cert. denied, 498 U.S. 1046 (1991).

U.S. v. An Article … Acu-Dot …, 483 F. Supp. 1311 (N.D. Ohio 1980).

U.S. v. Cecil, 836 F.2d 1431 (4th Cir. 1988), cert. denied, 487 U.S. 1205.

U.S. v. City of Chicago, 385 F. Supp. 543 (N.D. Ill. 1974).

U.S. v. De Gross, 960 F.2d 1433 (9th Cir. 1992).

U.S. v. Georgia Pacific Co., 474 F.2d 906 (5th Cir. 1973).

U.S. v. Gigante, 94 F.3d 53 (2nd Cir. 1996).

U.S. v. Goff, 509 F.2d 825 (5th Cir. 1975), cert. denied, 423 U.S.857 (1975).

U.S. v. Haldeman, 559 F.2d 31 (D.C. Cir., 1976), cert. denied, 431 U.S.933 (1977).

U.S. v. Hilton, 894 F.2d 485 (1st Cir. 1990).

U.S. v. Ironworkers Local 86, 443 F.2d 544 (9th Cir. 1971).

U.S. v. Jakobetz, 747 F. Supp. 250 (D. Vt. 1990).

U.S. v. Johnson, 990 F.2d. 1129 (9th Cir. 1993).

U.S. v. Kennedy, 548 F.2d. 608 (5th Cir. 1977), cert. denied, 434 U.S.865.

U.S. v. Kilgus, 571 F.2d 508 (9th Cir. 1978).

U.S. v. Lewis, 472 F.2d 252 (3rd Cir. 1973).

U.S. v. Nobles, 422 U.S. 225 (1975).

U.S. v. Northside Realty Assoc., 659 F.2d 590 (5th Cir. 1981) reversing 510 Fed. Supp. 668 (N.D. Ga. 1981).

U.S. v. Ortiz, 897 F. Supp. 199 (E.D. Pa. 1995).

U.S. v. Rodriguez, 588 F.2d 1003 (5th Cir. 1979).

U.S. v. Sgro, 816 F.2d 30 (1st Cir. 1981).

U.S. v. Shonubi, 802 F. Supp. 859 (E.D. N.Y. 1992), 998 F.2d 84 (2nd Cir. 1993), on remand, 895 F. Supp. 460 (E.D. N.Y. 1995), 103 F.3d 1085 (2nd Cir. 1997).

U.S. v. Skodnek, 933 F. Supp. 1108 (D. Mass. 1996).

U.S. v. Solomon, 753 F.2d 1522 (9th Cir. 1985).

U.S. v. Test, 550 F.2d 577 (10th Cir. 1976).

U.S. v. Two Bulls, 918 F.2d 56 (8th Cir. 1990).

U.S. v. U.S. Gypsum Co., 333 U.S. 364 (1948).

Valentino v. U.S. Postal Service, 674 F.2d 56 (D.C. Cir. 1982).

Virginia and Southern R.R. Co. v. Hawk, 160 F. 348 (1908), 87 C.C.A. 300.

Vuyanich v. Republic National Bank, 505 F. Supp. 2224 (N.D. Tex. 1980).

Wade v. New York Telephone Co., 500 F. Supp. 1170 (S.D. N.Y. 1980).

Ward's Cove Packing Co. v. Antonio, 490 U.S. 642 (1989).

Washington v. Davis, 426 U.S. 229 (1976).

Whitus v. Georgia, 385 U.S. 545 (1967).

Wilkins v. University of Houston, 654 F.2d 388, (5th Cir. 1981), vacated, 459 U.S. 809 (1982), on remand, 695 F.2d 134 (1983).

Williams et al. v. General Motors Corp., 656 F.2d 120 (5th Cir. 1981).

Williams v. City and County of San Francisco, 483 F. Supp. 335 (N.D. Cal. 1979).

Williams v. Superior Court, 49 Cal.3d 736, 256 Cal. Rptr. 503 (1989).

Wilmore v. City of Wilmington, 533 F. Supp. 844 (D. Del. 1982).

Winan v. New York & Erie R.R., 62 U.S. 88 (1853).

Windsurfing Int'l v. Fred Osterman GmbH, 613 F. Supp. 933 (S.D. N.Y. 1985).

Wisdom v. State, 234 Ga. 650, 217 S.E. 2d 244 (1975).

Witherspoon v. Illinois, 391 U.S. 510 (1968).

Yick Wo v. Hopkins, 118 U.S. 356 (1886).

Yorktown Medical Lab. Inc. v. Perales, 948 F.2d 84 (2nd Cir. 1991).

Zippo Manufacturing v. Rogers Imports, 216 F. Supp. 670 (S.D. N.Y. 1963).

Subject Index

Milton Keynes UK
Ingram Content Group UK Ltd.
UKHW040446071024
449327UK00020B/1029